FRANCIS CRICK

The
Astonishing Hypothesis
The Scientific Search
for the Soul

A Touchstone Book
Published by Simon & Schuster
New York London Toronto Sydney

TOUCHSTONE
Rockefeller Center
1230 Avenue of the Americas
New York, NY 10020

First Touchstone Edition 1995

TOUCHSTONE and colophon are registered trademarks
of Simon & Schuster Inc.

DESIGNED BY ERICH HOBBING

Manufactured in the United States of America

10

Library of Congress Cataloging-in-Publication Data
Crick, Francis, date.
    The Astonishing hypothesis: the scientific search for the soul/Francis Crick.
        p.   cm.
    Includes bibliographical references and index.
    1. Consciousness.   2. Mind and body.   3. Visual perception.   4. Neural circuitry.   I. Title.
BF311.C745   1993
153—dc20   92-46904
CIP
    ISBN 0-684-80158-2

For Christof Koch,
without whose energy and enthusiasm
this book would never have been written.

"**consciousness.** The having of perceptions, thoughts, and feelings; awareness. The term is impossible to define except in terms that are unintelligible without a grasp of what consciousness means. Many fall into the trap of equating consciousness with self-consciousness—to be conscious it is only necessary to be aware of the external world. Consciousness is a fascinating but elusive phenomenon; it is impossible to specify what it is, what it does, or why it evolved. Nothing worth reading has been written on it."

—Stuart Sutherland,
The International Dictionary
of Psychology

"As recently as a few years ago, if one raised the subject of consciousness in cognitive science discussions, it was generally regarded as a form of bad taste, and graduate students, who are always attuned to the social mores of their disciplines, would roll their eyes at the ceiling and assume expressions of mild disgust."

—John Searle

Contents

Preface

This book is about the mystery of consciousness—how to explain it in scientific terms. I do not suggest a crisp solution to the problem. I wish I could, but at the present time this seems far too difficult. Of course some philosophers are under the delusion they have already solved the mystery, but to me their explanations do not have the ring of scientific truth. What I have tried to do here is to sketch the general nature of consciousness and to make some tentative suggestions about how to study it experimentally. I am proposing a particular research strategy, not a fully developed theory. What I want to know is exactly what is going on in my brain when I see something.

Some readers will find this approach disappointing since, as a matter of tactics, it deliberately leaves out many aspects of consciousness they would love to hear discussed—in particular, how one should define it. You do not win battles by debating exactly what is meant by the word *battle*. You need to have good troops, good weapons, a good strategy, and then hit the enemy hard. The same applies to solving a difficult scientific problem.

I have tried to write for the general reader who has some interest in science but little expert knowledge of it. This means that I have had to explain the different disciplines related to consciousness in rather simple terms. Even so, some readers may find certain parts of the book difficult to read. To them I say: Don't be discouraged by the unfamiliar nature of some of the arguments or by the complexity of some of the experimental details. Press on, or skim over the difficult sections. The bottom line is often quite easy to understand.

Those philosophers and scientists who study the mind and the brain will see only too clearly that I have passed over many topics of vital interest to them. In spite of the simplicity of the treatment, I hope that they may learn something from what I have written, even if it is only in those sections they know least about. I have tried to avoid distortions of fact, although in biology this is not easy, if only because of the great variety of Nature. Distortions of viewpoint I cannot excuse so easily. Consciousness is a subject about which there is little consensus, even as to what the problem is. Without a few initial prejudices one cannot get anywhere. It will be clear to the reader that I am not at the moment enthusiastic about the views of functionalists, behaviorists, and some physicists, mathematicians, and philosophers. Tomorrow I may see (or be persuaded of) errors in my present thinking, but today I have to do the best I can.

The message of the book is that now is the time to think scientifically about consciousness (and its relationship, if any, to the hypothetical immortal soul) and, most important of all, the time to start the *experimental* study of consciousness in a serious and deliberate way.

To guide the reader through the jungle of the brain sciences the following summary of the book may be useful. It is in three main parts.

Chapter 1 starts with a bold statement about the Astonishing Hypothesis that encapsulates my approach to the brain—that to understand ourselves we must understand how nerve cells behave and how they interact. Earlier, prescientific ideas about consciousness and the soul are contrasted with our modern scientific knowledge of the universe. I then discuss briefly a few somewhat philosophical issues, such as reductionism, qualia, emergent behavior, and the reality of the world.

Chapter 2 outlines the general nature of consciousness (as described by William James a century ago and by three modern psychologists) and links it to attention and very short term memory. I then state the assumptions I make (and the attitudes I take) to come to grips with the problem and why I concentrate on one particular sort of consciousness—visual awareness—rather than on other kinds, such as consciousness of pain, self-consciousness, and so on.

Chapter 3 describes how the rather naïve ideas most people have about seeing are largely incorrect. Although we do not yet know exactly what happens in our brains when we see something, we can at least outline a possible way to approach the problem scientifically. Chapters 4 and 5, while fairly long, deal with only a few of the complexities of visual psychology. They should give the reader some idea of what has to be explained.

Part II is mainly a much oversimplified account of the brain and, in particular, its visual system. I have tried not to overwhelm the reader with too much detail while still providing enough information about the general way the nervous system is organized and how it works. I first describe the anatomy of the brain in outline (Chapter 7), followed by a simple account of single nerve cells (Chapter 8). Chapter 9 describes some experimental methods used to study the brain, its nerve cells, and its molecules. The next two chapters sketch the general nature of the visual system of higher primates. Chapter 12 illustrates how useful information can come from the study of human patients with damaged brains. Part II concludes with Chapter 13, which describes various theoretical models (called "neural networks") that simulate the behavior of small groups of neuronlike units.

The information in Parts I and II provides the necessary background to Part III, which deals with various possible experimental approaches to the problem of visual awareness. None of these has, as yet, led to a solution of the mystery, but several approaches look promising. Part III concludes with Chapter 18, which discusses some general issues that arise from my suggestions. This is followed by a short, informal postscript on Free Will.

To keep things tightly focused I have used footnotes for material not essential to the argument. I have also provided a glossary that briefly describes most of the scientific terms used in the text. This is prefaced by a short note on the names of the common scientific units of length, time, and frequency, since much of the action of the brain takes place at distances and times rather small compared to those of our everyday experience.

For readers wishing to follow up any topic, I have provided a list of books for further reading, some suitable for the lay reader and some for the expert. In most cases I have added a brief note about their content. The numbers in the text refer to a list of more technical references, published mainly in learned journals. These cover only a tiny fraction of the relevant scientific literature but should provide a toehold for further detailed exploration. I do not recommend these papers to the lay reader, if only because most of them are so badly written. There is no form of prose more difficult to understand and more tedious to read than the average scientific paper.

I shall be most grateful to readers who write to me to point out factual mistakes. I am less enthusiastic about corresponding on more general matters. Most people have their own ideas about consciousness and not a few feel compelled to put them on paper. I hope I may be forgiven for not reading everything sent to me on the subject. My normal practice is to consider only those ideas that have already been put forward in a refereed journal or in a book published by a reputable publisher. Otherwise the constant clatter of other people's suggestions makes it impossible for me to think effectively. I am still groping with these difficult problems, but I hope the reader will find this introduction to them of some interest.

Part I

"Men ought to know that from nothing else but the brain come joys, delights, laughter and sports, and sorrows, griefs, despondency and lamentations."

—Hippocrates (460–370 B.C.)

1

Introduction

Q: What is the soul?
A: The soul is a living being* without a body, having rea-
son and free will.
—Roman Catholic catechism

The Astonishing Hypothesis is that "You," your joys and your
sorrows, your memories and your ambitions, your sense of personal identity and free will, are in fact no more than the behavior of a vast assembly of nerve cells and their associated molecules. As
Lewis Carroll's Alice might have phrased it: "You're nothing but a
pack of neurons."† This hypothesis is so alien to the ideas of most
people alive today that it can truly be called astonishing.

The interest of human beings in the nature of the world, and about
their own natures in particular, is found in one form or another in all
peoples and tribes, however primitive. It goes back to the earliest
times from which we have written records and almost certainly from
before that, to judge from the widespread occurrence of careful
human burial. Most religions hold that some kind of spirit exists that

---

*As a small child my wife, Odile, was taught the catechism by an elderly Irish lady
who pronounced "being" as "be-in'." Odile heard this as "bean." She was extremely
puzzled by the idea of the soul as a living bean without a body but kept her worries to
herself.
†"Neuron" is the scientific word for a nerve cell.

3

persists after one's bodily death and, to some degree, embodies the essence of that human being. Without its spirit a body cannot function normally, if at all. When a person dies his soul leaves the body, although what happens after that—whether the soul goes to heaven, hell, or purgatory or alternatively is reincarnated in a donkey or a mosquito—depends upon the particular religion. Not all the religions agree in detail, but this is usually because they are based on different revelations—contrast the Christian Bible and the Muslim Koran. In spite of differences among religions, there is broad agreement on at least one point: People have souls, in the literal and not merely the metaphorical sense. These beliefs are held, and in many cases held strongly and aggressively, by the majority of human beings alive today.

There are, of course, a few exceptions. At one point, a minority of some of the more extreme Christians (following Aristotle) doubted whether women had souls, or at least had souls of the same quality as men. Some religions, such as Judaism, put little emphasis on life after death. Religions differ as to whether animals have souls. An old joke suggests that philosophers (in spite of all their differences) fall broadly into two classes: Those who own dogs, it is said, are confident that dogs have souls; those who do not, deny this.

Yet a minority of people today (including a large number in the former Communist countries) is inclined to a totally different view. Such people believe that the idea of a soul, distinct from the body and not subject to our known scientific laws, is a myth. It is easy to see how such myths could have arisen. Indeed, without a detailed knowledge of the nature of matter and radiation, and of biological evolution, such myths appear only too plausible.

Why, then, should this basic concept of the soul be doubted? Surely if almost everyone believed it, this is, in itself, prima facie evidence for it. But then some four thousand years ago almost everyone believed the earth was flat. The main reason for this radical change of opinion is the spectacular advance of modern science. Most of the religious beliefs we have today originated in a time when the earth, while a small place by our standards, was then thought of as being very large, even though its exact extent was unknown. Any one person had direct knowledge of only a tiny part of it. It was not implausible to believe that this large earth was the center of the universe and that man occupied the leading place in it. The earth's origins seemed lost in the mists of time and yet the span of time thought to be involved, while it seemed long in terms of human experience, we now know to be ridiculously short. It was not implausible to believe the earth was less than ten thousand years old. We now know its true age is about 4.6 billion years. The stars seemed far away, fixed perhaps in the spherical firmament, but that the universe extended as far as it does—more than 10 billion light years—was almost inconceivable. (An exception has to be made here for certain Eastern religions, such as Hinduism, that take pleasure in inflating times and distances for the sheer joy of it.)

Before Galileo and Newton our knowledge of basic physics was primitive. The movements of the sun and the planets seemed to be regular in some very intricate way. It was not totally unreasonable to believe that they needed angels to guide them. What else could make their behavior so regular? Even in the sixteenth and seventeenth centuries our understanding of chemistry was largely incorrect. Indeed, the existence of atoms was doubted by some physicists as late as the beginning of the twentieth century.

Yet today we know a great deal about the properties of atoms. We can give each type of chemical atom a unique whole number. We know their structure in detail, and most of the laws that control their behavior. Physics has provided a framework of explanation for chemistry. Our detailed knowledge of organic chemical molecules is enormous and expanding every day.

Admittedly, we still do not understand exactly what happens at very small distances (within the nucleus of an atom), at extremely high energies, and in very large gravitational fields. But for the conditions we normally have to deal with on earth (when an atom will only change into another atom under very special circumstances), most scientists think that this incompleteness of our knowledge probably matters very little, if at all, in our attempts to understand the mind and the brain.

In addition to our knowledge of basic chemistry and physics, the earth sciences (such as geology) and cosmic science (astronomy and cosmology) have developed pictures of our world and our universe that are quite different from those common when the traditional religions were founded. The modern picture of the universe, and how it developed in time, forms an essential background to our present knowledge of biology. That knowledge has been completely transformed in the last 150 years. Until Charles Darwin and Alfred Wallace independently hit on the basic mechanism driving biological evolution—the process of natural selection—the "Argument from Design" appeared to be unanswerable. How could an organism as complex and well designed as man have arisen without the help of an all-wise Designer? Yet this argument has collapsed completely. We now know that all living things, from bacteria to ourselves, are closely related at the biochemical level. We know that life has existed on earth for billions of years, and that the many species of plants and animals have changed, often radically, over that time. The dinosaurs have gone and in their place many new species of mammals have arisen. We can watch the basic processes of evolution happening today, both in the field and in our test tubes.

In this century there has been an equally dramatic biological advance, due to our understanding of the molecular nature of genes, the processes involved in their exact replication, together with our detailed knowledge of proteins and the mechanisms of their synthesis. We now realize that proteins, as a class, are immensely powerful and versatile, and can form the basis of elaborate biochemical devices. A major assault is being mounted on embryology (now often called "developmental biology"). The fertilized egg of a sea urchin normally divides many times, eventually turning into a mature sea urchin. If, after the first division, the two daughter cells of the fertilized egg are separated, then each of them will develop, in time, into a separate though rather smaller sea urchin. A similar experiment can be done on a frog's egg. The molecules have reorganized themselves to produce two little animals from the material that would otherwise have produced one. When this was first discovered, about a hundred years ago, it was suggested that some kind of immaterial Life Force must surely be at work. It seemed inconceivable that this dramatic doubling of a living creature could ever be explained on a biochemical basis—that is, using the properties of organic and other molecules and their interactions. Nowadays we feel that we shall have no difficulty in principle in working out how this can happen, although we expect the explanation to be complex. The history of science is littered with statements that something was inherently impossible to understand ("we shall never know of what the stars are made"). In many cases time has shown these predictions to be incorrect.

A modern neurobiologist sees no need for the religious concept of a soul to explain the behavior of humans and other animals. One is reminded of the question Napoleon asked after Pierre-Simon Laplace had explained to him the workings of the solar system: Where does God come into all this? To which Laplace replied, "Sire, I have no need of that hypothesis." Not all neuroscientists believe that the idea of the soul is a myth—Sir John Eccles[1,2] is the most notable exception—but certainly the majority do. It is not that they can yet prove the idea to be false. Rather, as things stand at the moment, they see no need for that hypothesis. Looked at in the perspective of human history, the main object of scientific research on the brain is not merely to understand and cure various medical conditions, important though this task may be, but to grasp the true nature of the human soul. Whether this term is metaphorical or literal is exactly what we are trying to discover.

Many educated people, especially in the Western world, also share the belief that the soul is a metaphor and that there is no personal life either before conception or after death. They may call themselves atheists, agnostics, humanists, or just lapsed believers, but they all deny the major claims of the traditional religions. Yet this does not mean that they normally think of themselves in a radically different way. The old habits of thought die hard. A man may, in religious terms, be an unbeliever but psychologically he may continue to think of himself in much the same way as a believer does, at least for everyday matters.

We need, therefore, to state the idea in stronger terms. The scientific belief is that our minds—the behavior of our brains—can be explained by the interactions of nerve cells (and other cells) and the molecules associated with them.* This is to most people a really surprising concept. It does not come easily to believe that I am the detailed behavior of a set of nerve cells, however many there may be and however intricate their interactions. Try for a moment to imagine this point of view. ("Whatever he may say, Mabel, I know I'm in there somewhere, looking out on the world.")

Why does the Astonishing Hypothesis seem so surprising? I think there are three main reasons. The first is that many people are reluctant to accept what is often called the "reductionist approach"—that a complex system can be explained by the behavior of its parts and their interactions with each other. For a system with many levels of activity, this process may have to be repeated more than once—that is, the behavior of a particular part may have to be explained by the properties of *its* parts and their interactions. For example, to understand the brain we may need to know the many interactions of nerve cells with each other; in addition, the behavior of each nerve cell may need explanation in terms of the ions and molecules of which it is composed.

Where does this process end? Fortunately there is a natural stop-

---

*This idea is not novel. An especially clear statement of it can be found in a well-known paper by Horace Barlow.[3]

ping point. This is at the level of the chemical atoms. Each atom consists of a heavy atomic nucleus, carrying a positive charge, surrounded by an organized cloud of light, negatively charged nimble electrons. The chemical properties of each atom are determined almost entirely by its nuclear charge.* The other properties of the nucleus—its mass, its secondary electrical properties such as the strengths of its dipole, and its quadrupole—make in most cases only small differences to its chemical properties.

Now, to a first approximation, the mass and charge of the nucleus of an atom never change, at least in the mild environment in which life flourishes on earth. Thus the knowledge of the substructure of the nucleus is not needed for chemistry. It makes no difference that an atomic nucleus is composed of various combinations of protons and neutrons, and that they, in turn, are made up of quarks. All the chemist needs to know about each atom is its nuclear charge in order to explain most of the facts of chemistry. To do this he needs to understand the rather unexpected type of mechanics (called "quantum mechanics") that controls the behavior of very small particles and of electrons in particular. In practice, since the calculations soon become impossibly intricate, he mainly uses various rules-of-thumb that we now can see have a reasonable explanation in terms of quantum mechanics. Below this level he need not venture.[†]

There have been a number of attempts to show that reductionism cannot work. They usually take the form of a rather formal definition, followed by an argument that reductionism of this type cannot be true. What is ignored is that reductionism is not the rigid process of explaining one fixed set of ideas in terms of another fixed set of ideas at a lower level, but a dynamic interactive process that modifies the concepts at both levels as knowledge develops. After all, "reductionism" is the main theoretical method that has driven the development of physics, chemistry, and molecular biology. It is largely responsible for the spectacular developments of modern science. It is the only sensible way to proceed until and unless we are confronted with strong

---

*A carbon nucleus has a charge of +6; an oxygen nucleus a charge of +8. Thus an oxygen atom, to be electrically neutral, must have 8 negatively charged electrons associated with it.

†The major exception to all this is radioactivity: the rare change of one atom into another that occurs in stars, atomic piles and bombs, and, less spectacularly, in the atoms of radioactive minerals and in specially contrived experiments in the laboratory. Radioactivity can produce mutations in DNA, the genetic material, so it cannot be ignored completely, but it is unlikely to be important as a basic process in the behavior of our brains.

experimental evidence that demands we modify our attitude. General philosophical arguments against reductionism will not do.

Another favorite philosophical argument is that reductionism involves a "category mistake." For example, in the 1920s this could have taken the form that to consider a gene to be a molecule (or as we should say now, part of a matched pair of molecules) would be a category mistake. A gene is one category and a molecule is a quite different category. One can see now how hollow such objections have turned out to be.* Categories are not given to us as absolutes. They are human inventions. History has shown that although a category may sound very plausible, it can, on occasion, turn out to be both misconceived and misleading. Recall the four Humors in ancient and medieval medicine: blood, phlegm, choler, and black bile.

The second reason why the Astonishing Hypothesis seems so strange is the nature of consciousness. We have, for example, a vivid internal picture of the external world. It might seem a category mistake to believe this is merely another way of talking about the behavior of neurons, but we have just seen that arguments of this type are not always to be trusted.

Philosophers have been especially concerned with the problem of qualia—for example, how to explain the redness of red or the painfulness of pain. This is a very thorny issue. The problem springs from the fact that the redness of red that I perceive so vividly cannot be precisely communicated to another human being, at least in the ordinary course of events. If you cannot describe the properties of a thing unambiguously, you are likely to have some difficulty trying to explain those properties in reductionist terms. This does not mean that, in the fullness of time, it may not be possible to explain to you the *neural correlate*† of your seeing red. In other words, we may be able to say that you perceive red if and only if certain neurons (and/or molecules) in your head behave in a certain way. This may, or may not, suggest *why* you experience the vivid sensation of color and why one sort of neural behavior necessarily makes you see red while another makes you see blue, rather than vice versa.

Even if it turns out that the redness of red cannot be explained (because you cannot communicate that redness to me), it does not follow that we cannot be reasonably sure that you see red in the same

---

*The Canadian philosophers Paul and Patricia Churchland (now at the University of California, San Diego) have dealt very satisfactorily with the arguments against reductionism—see the section on further reading for references to their work.

†This useful term is worth committing to memory.

way as I see red. If it turns out that the neural correlate of red is exactly the same in your brain as in mine, it would be scientifically plausible to infer that you see red as I do. The problem lies in the word "exactly." How precise we have to be will depend on a detailed knowledge of the processes involved. If the neural correlate of red depends, in an important way, on my past experience, and if my past experience is significantly different from yours, then we may not be able to deduce that we both see red in exactly the same way.

One may conclude, then, that to understand the various forms of consciousness we first need to know their neural correlates.

The third reason why the Astonishing Hypothesis seems strange springs from our undeniable feeling that our Will is free. Two problems immediately arise: Can we find a neural correlate of events we consider to show the free exercise of our Will? And could it not be that our Will only appears to be free? I believe that if we first solve the problem of awareness* (or consciousness), the explanation of Free Will is likely to be easier to solve. (This topic is discussed at greater length in the Postscript on page 265.)

How did this extraordinary neuronal machine arise? To understand the brain, it is important to grasp that it is the end product of a long process of evolution by natural selection. It has not been designed by an engineer, even though, as we shall see, it does a fantastic job in a small space and uses relatively little energy to do it. The genes we received from our parents have, over many millions of years, been influenced by the experience of our distant ancestors. These genes, and the processes directed by them before birth, lay down much of the structure of the parts of our brain. The brain at birth, we now know, is not a tabula rasa but an elaborate structure with many of its parts already in place. Experience then tunes this rough-and-ready apparatus until it can do a precision job.

Evolution is not a clean designer. Indeed, as François Jacob, the French molecular biologist, has written. "Evolution is a tinkerer."[4] It builds, mainly in a series of smallish steps, on what was there before. It is opportunistic. If a new device works, in however odd a manner, evolu-

---

*I have used the terms *awareness* and *consciousness* more or less interchangeably, although I tend to use awareness (as in "visual awareness") for some particular aspect of consciousness. Some philosophers make a distinction between them but there is no general agreement as to how such a distinction should be made. I must confess that in conversation I find I say "consciousness" when I want to startle people and "awareness" when I am trying not to.

tion will try to promote it. This means that changes and improvements that can be added to the existing structures with relative ease are more likely to be selected, so the final design may not be a clean one but rather a messy accumulation of interacting gadgets. Surprisingly, such a system often works better than a more straightforward mechanism that is designed to do the job in a more direct manner.

Thus the mature brain is the product of both Nature and Nurture. We can see this easily in the case of language. The ability to handle a complex language fluently appears to be unique to human beings. Our nearest relatives, the apes, perform very poorly at language use even after extensive training. Yet the actual language we learn is obviously heavily dependent on where and how we were brought up.

Two more philosophical points need to be made. The first is that much of the behavior of the brain is "emergent"—that is, behavior does not exist in its separate parts, such as the individual neurons. An individual neuron is in fact rather dumb. It is the intricate interaction of many of them together that can do such marvelous things.

There are two meanings of the term *emergent*. The first has mystical overtones. It implies that the emergent behavior cannot in any way, even in principle, be understood as the combined behavior of its separate parts. I find it difficult to relate to this type of thinking. The scientific meaning of emergent, or at least the one I use, assumes that, while the whole may not be the simple sum of the separate parts, its behavior can, at least in principle, be *understood* from the nature and behavior of its parts *plus* the knowledge of how all these parts interact.

A simple example, from elementary chemistry, would be any organic compound, such as benzene. A benzene molecule is made of six carbon atoms arranged symmetrically in a ring with a hydrogen atom attached, on the outside of the ring, to each carbon atom. Apart from its mass, the properties of a benzene molecule are not in any sense the simple arithmetical sum of the properties of its twelve constituent atoms. Nevertheless, the behavior of benzene, such as its chemical reactivity and its absorption of light, can be calculated if we know how these parts interact, although we need quantum mechanics to tell us how to do this. It is curious that nobody derives some kind of mystical satisfaction by saying "the benzene molecule is more than the sum of its parts," whereas too many people are happy to make such statements about the brain and nod their heads wisely as they do so. The brain is so complicated, and each brain is so individual, that we may never be able to obtain second-to-second detailed knowledge of how a particular brain works but we may hope at least to understand the general principles of how complex sensations and behaviors arise in the brain from the interactions of its many parts.

Of course there may be important processes going on that have not yet been discovered. I suspect that even if we were told the exact behavior of one part of the brain we might in some cases not immediately understand the explanation, since it might involve new concepts and new ideas that have yet to be articulated. However, I do not share the pessimism of some who think that our brains are inherently incapable of grasping such ideas. I prefer to confront such difficulties, if indeed they exist, when we come to them. Our brains have evolved and developed so that we can deal fluently with many concepts related to our everyday world. Nevertheless, well-trained brains can grasp ideas about phenomena that are not part of our normal experience, such as those of relativity and quantum mechanics. Such ideas are very counterintuitive, but constant practice with them enables the trained brain to grasp them and manipulate them easily. Ideas *about* our brains are likely to have the same general character. They may appear very strange at first but with practice we may hope to handle them with confidence.

There is no obvious reason why we should not be able to obtain this knowledge—both of the components of the brain and also how they interact together. It is the sheer variety and complexity of the processes involved that makes our progress so slow.

The second philosophical conundrum that needs clarification concerns the reality of the outside world. Our brains have evolved mainly to deal with our body and its interactions with the world it senses to be around us. Is this world real? This is a venerable philosophical issue and I do not wish to be embroiled in the finely honed squabbles to which it has led. I merely state my own working hypotheses: that there is indeed an outside world, and that it is largely independent of our observing it. We can never fully know this outside world, but we can obtain approximate information about some aspects of its properties by using our senses and the operations of our brain. Nor, as we shall see, are we aware of everything that goes on in our brains, but only of some aspects of that activity. Moreover, both these processes—our interpretations of the nature of the outside world and of our own introspections—are open to error. We may think we know our motives for a particular action, but it is easy to show that, in some cases at least, we are in fact deceiving ourselves.

2

The General Nature
of Consciousness

"In any field find the strangest thing and then explore it."
—John Archibald Wheeler

To come to grips with the problem of consciousness, we first need to know what we have to explain. Of course, in a general way we all know what consciousness is like. Unfortunately that is not enough. Psychologists have frequently shown that our common-sense ideas about the workings of the mind can be misleading. The obvious first step, then, is to find out what psychologists over the years have considered to be the essential features of consciousness. They may not have got it quite right, but at least their ideas on the subject will provide us with a starting point.

Since the problem of consciousness is such a central one, and since consciousness appears so mysterious, one might have expected that psychologists and neuroscientists would now direct major efforts toward understanding it. This, however, is far from being the case. The majority of modern psychologists omit any mention of the problem, although much of what they study enters into consciousness. Most modern neuroscientists ignore it.

This was not always so. When psychology began as an experimental science, largely in the latter part of the nineteenth century, there was

13

much interest in consciousness, even though it was admitted that the exact meaning of the word was unclear. The major method of studying it, especially in Germany, was by detailed and systematic introspection. It was hoped that psychology might become more scientific by refining introspection until it became a reliable technique.

The American psychologist William James (the brother of novelist Henry James) discussed consciousness at some length. In his monumental work *The Principles of Psychology*, first published in 1890, he described five properties of what he called "thought." Every thought, he wrote, tends to be part of personal consciousness. Thought is always changing, is sensibly continuous, and appears to deal with objects independent of itself. In addition, thought focuses on some objects to the exclusion of others. In other words, it involves attention. Of attention he wrote, in an oft-quoted passage: "Everyone knows what attention is. It is the taking possession by the mind, in clear and vivid form, of one out of what seem several simultaneously possible objects or trains of thought. . . . It implies withdrawal from some things in order to deal effectively with others."

In the nineteenth century we can also find the idea that consciousness is closely associated with memory. James quotes the Frenchman Charles Richet, who wrote in 1884 that "to suffer for only a hundredth of a second is not to suffer at all; and for my part I would readily agree to undergo a pain, however acute and intense it might be, provided it should last only a hundredth of a second, and leave after it neither reverberation nor recall."

Not all operations of the brain were thought to be conscious. Many psychologists believed that some processes are subliminal or subconscious. For example, Hermann von Helmholtz, the nineteenth-century German physicist and physiologist, often spoke of perception in terms of "unconscious inference." By this he meant that perception was similar in its logical structure to what we normally mean by inference, but that it was largely unconscious.

In the early twentieth century the concepts of the preconscious and the unconscious were made widely popular, especially in literary circles, by Freud, Jung, and their associates, mainly because of the sexual flavor they gave to them. By modern standards, Freud can hardly be regarded as a scientist but rather as a physician who had many novel ideas and who wrote persuasively and unusually well. He became the main founder of the new cult of psychoanalysis.

Thus, as long as one hundred years ago we see that three basic ideas were already current:

1. Not all the operations of the brain correspond to consciousness.
2. Consciousness involves some form of memory, probably a very short term one.
3. Consciousness is closely associated with attention.

Unfortunately, a movement arose in academic psychology that denied the usefulness of consciousness as a psychological concept. This was partly because experiments involving introspection did not appear to be leading anywhere and partly because it was hoped that psychology could become more scientific by studying behavior (in particular, animal behavior) that could be observed unambiguously by the experimenter. This was the Behaviorist movement. It became taboo to talk about mental events. All behavior had to be explained in terms of the stimulus and the response.

Behaviorism was especially strong in the United States, where it was started by John B. Watson and others before World War I. It flourished in the 1930s and 1940s when B. F. Skinner was its most celebrated exponent. Other schools of psychology existed in Europe, such as the Gestalt school (discussed in Chapter 4), but it was not until the rise of cognitive psychology in the late 1950s and 1960s that it became intellectually respectable, at least in the United States, for a psychologist to talk about mental events. It then became possible to study visual imagery[1] for example, and to postulate psychological models for various mental processes, usually based on concepts used to describe the behavior of digital computers. Even so, consciousness was seldom mentioned and there was little attempt to distinguish between conscious and unconscious activity in the brain.

Much the same was true of neuroscientists studying the brains of experimental animals. Neuroanatomists worked almost entirely on dead animals (including human beings) while neurophysiologists largely studied anesthetized animals who are not conscious. For example, they do not feel any pain during such experiments. This was especially true after the epoch-making discovery by the neurobiologists David Hubel and Torsten Wiesel in the late fifties. They found that nerve cells in the visual cortex of the brain of an anesthetized cat showed a whole series of interesting responses when light was shone on the cat's opened eyes, even though its brain waves showed it to be more asleep than awake. For this and subsequent work they were awarded a Nobel Prize in 1981.

It is far more difficult to study the response of these brain cells in an alert animal (not only must its head be restrained but eye movements must either be prevented or carefully observed). For this reason very few experiments have been done to compare the responses of the same brain cell to the same visual signals under two conditions: when the animal is asleep and again when it is awake. Not only have neuroscientists traditionally avoided the problem of consciousness because of these experimental difficulties but also because they considered the problem both too subjective and too "philosophical," and thus not easily amenable to experimental study. It would not have been easy for a neuroscientist to get a grant for the purpose of studying consciousness.

Physiologists are still reluctant to worry about consciousness, but in the last few years a number of psychologists have begun to address the matter. I will sketch briefly the ideas of three of them. What they have in common is a neglect or, at best, a distant interest in nerve cells. Instead, they hope to contribute to an understanding of consciousness mainly by using the standard methods of psychology. They treat the brain as an impenetrable "black box" of which we know only the outputs—the behaviors it produces—caused by various inputs, such as those signalled by the senses. And they construct models that use general ideas based on our commonsense understanding of the mind, which they express in engineering or computing terms. All three authors would probably describe themselves as cognitive scientists.

Philip Johnson-Laird, now a professor of psychology at Princeton University, is a distinguished British cognitive scientist with a major interest in language, especially in the meaning of words, sentences, and narratives. These are issues unique to humans. It is not surprising that Johnson-Laird pays little attention to the brain, since much of our detailed information about the primate brain is derived from monkeys and they have no true language. His two books, *Mental Models* and *The Computer and the Mind*, are concerned with the problem of how to describe the mind—the activities of the brain—and the relevance of modern computers to that description.[2,3] He stresses that the brain is, as we shall see, a highly parallel mechanism (meaning that millions of processes are going on at the same time) and that we are unconscious of much of what it does.*

Johnson-Laird believes that any computer, and especially a highly parallel computer, must have an operating system that controls the

---

*Johnson-Laird is especially interested in self-reflection and self-awareness, topics that, for tactical reasons, I shall leave to one side.

rest of its functions, even though it may not have complete control over them. He proposes that it is the workings of this operating system that correspond most closely to consciousness and that it is located at a high level in the hierarchies of the brain.

Ray Jackendoff, professor of linguistics and cognitive science at Brandeis University, is a well-known American cognitive scientist with a special interest in language and music. Like most cognitive scientists, he believes that the mind is best thought of as a biological information-processing system. However, he differs from many of them in that he regards, as one of the most fundamental issues of psychology, the question "What makes our conscious experience the way it is?"

His Intermediate-Level Theory of Consciousness states that awareness is derived neither from the raw elements of perception nor from high-level thought but from a level of representation intermediate between the most peripheral (sensationlike) and the most central (thoughtlike). He rightly emphasizes that this is a quite novel point of view.[4]

Like Johnson-Laird, Jackendoff has been strongly influenced by the analogy of the brain to a modern computer. He points out that this analogy provides some immediate dividends. For example, in a computer, a great deal of the information is stored but only a small part is active at any one time. The same is true of the brain.

However, not all of the activity in the brain is conscious. He thus makes a distinction not merely between the brain and the mind but between the brain, the computational mind, and what he calls the "phenomenological mind," meaning (roughly) what we are conscious of. He agrees with Johnson-Laird that what we are conscious of is the *result* of computations rather than the computations themselves.*

He also believes there is an intimate connection between awareness and short-term memory. He expresses this by saying that "awareness is supported by the contents of short-term memory," going on to add that short-term memory involves "fast" processes and that slow processes have no direct phenomenological effect.

As to attention, he suggests that the computational effect of attention is that the material being attended to is undergoing especially intensive and detailed processing. He believes that this is what accounts for attention's limited capacity.

Jackendoff and Johnson-Laird are both functionalists. Just as it is not necessary to know about the actual wiring of a computer when

---

*Jackendoff expresses this in jargon of his own. What I have called "the results" he calls "information structures."

writing programs for it, so a functionalist investigates the information processed by the brain, and the computational processes the brain performs on this information, without considering the neurological implementation of these processes. He usually regards such considerations as totally irrelevant or, at best, premature.*

This attitude does not help when one wants to *discover* the workings of an immensely complicated apparatus like the brain. Why not look inside the black box and observe how its components behave? It is not sensible to tackle a very difficult problem with one hand tied behind one's back. When, eventually, we know in some detail how the brain works, then a high-level description (which is what functionalism is) may be a useful way to think about its overall behavior. Such ideas can always be checked for accuracy by using detailed information from lower levels, such as the cellular or molecular levels. Our provisional high-level descriptions should be thought of as rough guides to help us unravel the intricate operations of the brain.

Bernard J. Baars, an institute faculty professor at the Wright Institute in Berkeley, California, has written a book entitled *A Cognitive Theory of Consciousness*.[5] Although Baars is a cognitive scientist, he is rather more interested in the human brain than either Jackendoff or Johnson-Laird.

He calls his basic idea the Global Workspace. He identifies the information that exists at any one time in this workspace as the content of consciousness. The workspace, which acts as a central informational exchange, is connected to many unconscious receiving processors. These specialists are highly efficient in their own domains but not outside them. In addition, they can cooperate and compete for access to the workspace. Baars elaborates this basic model in several ways. For example, the receiving processors can reduce uncertainty by interacting until they reach agreement on only one active interpretation.†

In broader terms, he considers consciousness to be profoundly active and that there are attentional control mechanisms for access to consciousness. He believes we are conscious of some items in short-term memory but not all.

---

*Genetics is also concerned with information transfer, both between generations and within an individual, but the real breakthrough came when the structure of DNA showed very clearly the idiom in which this information was expressed.

†I have not attempted to describe all the complications of Baars's model. Many of these have been added because he wishes to explain many aspects of consciousness, such as consciousness of self, self-monitoring, and also other psychological activities such as unconscious contexts, volition, hypnosis, and so on.

These three cognitive theoreticians share a loose consensus on three points about the nature of consciousness. They all agree that not all the activities of the brain correspond directly with consciousness, and that consciousness is an active process. They all believe that attention and some form of short-term memory is involved in consciousness. And they would probably agree that the information in consciousness can be fed into both long-term episodic memory as well as into the higher, planning levels of the motor system that controls voluntary movements. Beyond that their ideas diverge somewhat.

Let us keep all three sets of ideas in mind while exploring an approach that tries to see what we can learn by combining them with our growing knowledge of the structure and activity of the nerve cells in the brain.

Most of my own ideas on consciousness have been developed in collaboration with a younger colleague, Christof Koch, now a professor of computation and neural systems at the California Institute of Technology (Caltech). Christof and I have known each other from the time in the early eighties when he was a graduate student of Tomaso Poggio in Tübingen. Our approach is essentially a scientific one.* We believe that it is hopeless to try to solve the problems of consciousness by general philosophical arguments; what is needed are suggestions for new experiments that might throw light on these problems. To do this we need a tentative set of theoretical ideas that may have to be modified or discarded as we proceed. It is characteristic of a scientific approach that one does not try to construct some all-embracing theory that purports to explain *all* aspects of consciousness. Nor does such an approach concentrate on studying language simply because it is uniquely human. Rather, one tries to select the most favorable system for the study of consciousness, as it appears at this time, and study it from as many aspects as possible. In a battle, you do not usually attack on all fronts. You probe for the weakest place and then concentrate your efforts there.

We made two basic assumptions. The first is that there is something that requires a scientific explanation. There is general agreement that people are not conscious of all the processes going on in their heads, although exactly which might be a matter of dispute. While you are aware of many of the results of perceptual and memory

---

*In what follows, I quote extensively from ideas that Koch and I published on this subject in 1990 in the journal called *Seminars in the Neurosciences* (SIN).[6]

processes, you have only limited access to the processes that produce this awareness (e.g., "How did I come up with the first name of my grandfather?"). In fact, some psychologists have suggested that you have only very limited introspective access to the origins of even higher order cognitive processes. It seems probable, however, that at any one moment some active neuronal processes in your head correlate with consciousness, while others do not. *What are the differences between them?*

Our second assumption was tentative: that all the different aspects of consciousness, for example pain and visual awareness, employ a basic common mechanism or perhaps a few such mechanisms. If we could understand the mechanisms for one aspect, then we hope we will have gone most of the way to understanding them all. Paradoxically, consciousness appears to be so odd and, at first sight, so difficult to understand that only a rather special explanation is likely to work. The general nature of consciousness may be easier to discover than more mundane operations, such as how the brain processes information so that you see in three dimensions, which can, in principle, be explained in many different ways. Whether this is really true remains to be seen.

Christof and I suggested that several topics should be set aside or merely stated outright without further discussion, for experience has shown that otherwise much valuable time can be wasted arguing about them.

1. Everyone has a rough idea of what is meant by consciousness. It is better to avoid a *precise* definition of consciousness because of the dangers of premature definition. Until the problem is understood much better, any attempt at a formal definition is likely to be either misleading or overly restrictive, or both.*

2. Detailed arguments about what consciousness is for are probably premature, although such an approach may give useful hints about its nature. It is, after all, a bit surprising that one should worry too much about the function of something when we are rather vague about what it is. It is known that without consciousness you can deal only with familiar, rather routine, situations or respond to very limited information in new situations.

3. It is plausible that some species of animals—and in particular the higher mammals—possess some of the essential features of conscious-

*If this seems like cheating, try defining for me the word *gene*. So much is now known about genes that any simple definition is likely to be inadequate. How much more difficult, then, to define a biological term when rather little is known about it.

ness, but not necessarily all. For this reason, appropriate experiments on such animals may be relevant to finding the mechanisms underlying consciousness. It follows that a language system (of the type found in humans) is not essential for consciousness—that is, one can have the key features of consciousness without language. This is not to say that language may not considerably enrich consciousness.

4. It is not profitable at this stage to argue about whether "lower" animals, such as octopus, fruit flies, or nematodes, are conscious. It is probable, however, that consciousness correlates to some extent with the degree of complexity of any nervous system. When we clearly understand, both in detail and in principle, what consciousness involves in humans, then will be the time to consider the problem of consciousness in much lower animals.

For the same reason I won't ask whether some parts of our own nervous system have a special, isolated, consciousness of their own. If you say, "Of course my spinal cord is conscious but it's not telling me," I am not, at this stage, going to spend time arguing with you about it.

5. There are many forms of consciousness, such as those associated with seeing, thinking, emotion, pain, and so on. Self-consciousness— that is, the self-referential aspect of consciousness—is probably a special case of consciousness. In our view, it is better left to one side for the moment. Various rather unusual states, such as the hypnotic state, lucid dreaming, and sleep walking, will not be considered here since they do not seem to have special features that would make them experimentally advantageous.

How can we approach consciousness in a scientific manner? Consciousness takes many forms, but as I have already explained, for an initial scientific attack it usually pays to concentrate on the form that appears easiest to study. Christof Koch and I chose visual awareness rather than other forms of consciousness, such as pain or self-awareness, because humans are very visual animals and our visual awareness is especially vivid and rich in information. In addition, its input is often highly structured yet easy to control. For these reasons much experimental work has already been done on it.

The visual system has another advantage. There are many experiments that, for ethical reasons, cannot be done on humans but can be done on animals. (This is discussed more fully in Chapter 9.) Fortunately, the visual system of higher primates appears somewhat similar to our own, and many experiments on vision have already been done on animals such as the macaque monkey. If we had chosen to study the language system there would have been no suitable experimental animal to work on.

Because of our detailed knowledge of the visual system in the primate brain (discussed in Chapters 10 and 11), we can see how the visual parts of the brain take the picture (the visual field) apart, but we do not yet know how the brain puts it all together to provide our highly organized view of the world—that is, what we see. It seems as if the brain needs to impose some global unity on certain activities in its different parts so that the attributes of a single object—its shape, color, movement, location, and so on—are in some way brought together without at the same time confusing them with the attributes of other objects in the visual field.

This global process requires mechanisms that could well be described as "attention" and involves some form of very short term memory. It has been suggested that this global unity might be expressed by the *correlated* firing of the neurons involved. Loosely speaking, this means that the neurons that respond to the properties of that particular object tend to fire at the same moment, whereas other active neurons, corresponding to other objects, do not fire in synchrony with the correlated set. (This is discussed more fully in Chapters 14 and 17.) To approach the problem we must first understand something of the psychology of vision.

3

Seeing

"Seeing is believing."

I am often asked by nonscientists, usually at the dinner table, what I am working on. When I say I am thinking about some of the problems of the visual system of mammals—that is, how we see things—there is usually a slightly embarrassed pause. My questioner is wondering why there should be any difficulty about something as simple as seeing. After all, we open our eyes and there the world is, large and clear, full of objects in vivid Technicolor, without our having to make any appreciable effort. It all seems so delightfully easy, so what can be the problem? Now, if my preoccupation had been how we do mathematics or understand chemistry or (even worse) economics—any of which requires some mental effort—there might be something worth talking about. But seeing. . .?

In addition, many people feel that if their brain is working well, why should they bother about it. The main "problem" associated with human brains, they believe, is how to cure them when something has gone wrong. Only a few scientifically minded people go further and ask: How exactly does my brain work when I see something?

There are two rather surprising aspects of our present knowledge of the visual system. The first is how much we already know—by any standards the amount is enormous. Whole courses are given on the psychology of

23

vision (for example, under what conditions is apparently smooth movement produced by the rapid succession of still pictures on the cinema screen), the physiology of vision (the structure and behavior of the eye and the relevant parts of the brain), and the molecular and cell biology of vision (nerve cells and their many component molecules). This knowledge has been acquired by the painstaking efforts of many experimenters and theorists, over many years, studying both humans and animals.

The other surprising thing is that, in spite of all this work, we really have no clear idea how we see anything. This fact is usually concealed from the students who take such courses. Surely after all that careful work and all those elaborate arguments it would be bad form to suggest that we still lack any clear scientific understanding of the process of vision. And yet, by the standards of the exact sciences (such as physics, chemistry, and molecular biology), we do not yet know, even in outline, how our brains produce the vivid visual awareness that we take so much for granted. We can glimpse fragments of the processes involved, but we lack both the detailed information and the ideas to answer the most simple questions: How do I see color? What is happening when I recall the image of a familiar face? and so on.

But there is a third surprising thing. You probably already have a rough-and-ready idea of how you yourself see things. You know that each of your eyes is somewhat like a little television camera. Using a lens, it focuses the visual scene before you onto a special screen—the retina—at the back of each eye. Each retina has many millions of individual "photoreceptors" that respond to the photons—the particles of light—coming into your eye. Then "you" put together the pictures coming to your brain from your two eyes, and so you see. Without thinking about it, you probably have some idea how this might happen. What may surprise you is that, even if scientists still do not know how we see things, it is easy to show that how *you* think you see things is largely simplistic or, in many cases, plain wrong.

The mental picture that most of us have is that there is a little man (or woman) somewhere inside our brain who is following (or, at least, trying hard to follow) what is going on. I shall call this the Fallacy of the Homunculus (*homunculus* is Latin for "little man"). Many people do indeed feel this way—and that fact, in due course, will itself need an explanation—but our Astonishing Hypothesis states that this is not the case. Loosely speaking, it says that "it's all done by neurons."

Given this hypothesis, *the problem of seeing takes on a totally new character*. In short, there must be structures or operations in the brain that, in some mysterious way, behave as if they correspond somewhat to the mental picture of the homunculus. But what could they possibly be? To approach this difficult question, we have to know something about what the task of seeing involves and about the biological apparatus inside our heads that does the job.

Why do you need a visual system? The glib answer is so that you (or your genes) can leave more descendants, or to help your relatives leave more descendants, but this is too general to tell us much. In practice, an animal needs to see to obtain food, to avoid predators and other dangers, to mate, to raise its young (in some species), and so on. For this a good visual system is invaluable.

The neurobiologist John Allman, who works at Caltech, has argued that mammals, in contrast to reptiles, have a special need to preserve heat because of their continual activity and their relatively high and constant body temperature. This is especially true for small mammals, since their surface area is so big compared to their volume. Hence fur, a uniquely mammalian attribute, and also, he suggests, the large development of the mammalian neocortex. He believes that this part of the brain made the early mammals smart enough to locate sufficient food to keep them warm.

Although mammals are smart, they do not, as a class, have especially good vision, probably because they evolved from small nocturnal creatures for whom vision was less important than smell and hearing. The exceptions are the primates (monkeys, apes, and man), most of whom have evolved excellent vision, although, as in man, their sense of smell may be poor.

When the dinosaurs were all killed off, these early mammals evolved rapidly as they took over the ecological niches left empty by the dinosaurs. The smarter brains of the mammals helped them to do this very effectively, and eventually led to the emergence of man, the smartest mammal of them all.

So what do mammals use their eyes for? The photons coming into our eyes only show us how much light comes from each particular part of the visual field,* plus some information about its wavelength. But what you want to know is *what* is out there, what it is doing, and what

---

*A more correct term is *stimulus field* but I feel that terms such as *visual field*, *field of vision*, or *visual scene* sound better to most readers. It is of course important to distinguish between objects in the external world and the processes in your head that correspond to your seeing them.

it is likely to do. In other words, you need to see objects, their movements, and something about their "meaning"—what they usually do, or are usually used for, when, and under what circumstances you have seen them (or similar objects) in the past. . .and so on.

Not only do you need all this information if you are to survive and leave viable offspring, but you need it, in computer terminology, "in real time"—meaning sufficiently quickly that you can do something about it before it becomes stale. It is of little use to produce a highly accurate forecast of tomorrow's weather if it takes a week to calculate it. So there is a very high premium on extracting the vital information as quickly as possible. This is especially true when one animal is trying to kill another animal, both for the predator and for the prey.

Thus the eye and the brain must attempt to analyze the incoming light so that it provides all this important information. How does it do this? Before I describe in more detail exactly what is involved in seeing, let me make three general remarks.

1. You are easily deceived by your visual system.
2. The visual information provided by your eyes can be ambiguous.
3. Seeing is a constructive process.

Let us take these in turn, although they are not unrelated.

You are easily deceived by your visual system. For example, many people believe they see everything before them with equal clarity. As I look out my study window onto the garden, I have the impression that I see the rose bushes ahead of me just as clearly as the trees that are farther to my right. If I keep my eyes still for a moment I can easily observe that this is false. I can see fine details only when they are close to my center of gaze. Off to one side of it my vision becomes increasingly blurred. At the extreme periphery of vision I can identify objects only with difficulty. These limitations are not immediately apparent in everyday life because we move our eyes easily and frequently so that we have the *illusion* of seeing equally clearly everywhere.

Hold a colored object, such as a blue pen or a red playing card, to one side of your head and sufficiently far back that you cannot see it at all. Gradually bring it forward so that it just enters the periphery of your field of view. BE SURE NOT TO MOVE YOUR EYES. If you jiggle the object, you will recognize that something is moving there before you can see what it is. You can tell if the pen (or the playing card) is horizontal or vertical before you can be sure what color it is. Even when you can see both the shape and the color you still will not see the fine details of the object until you bring it quite close to the center of gaze. My pen has a little label on it that says "extra fine point." The print is small but with my glasses on I can easily read it if I hold it about a foot from my eyes. But if I put my finger next to it, and gaze not at the pen but at the center of my fingertip, I cannot read what is written on the pen, even though the writing is only a short distance from my center of gaze. My visual acuity, as it is called, falls off very rapidly away from the center of gaze.

To demonstrate, in a simple direct way, how the visual system can fool you, look at Figure 1. You immediately see this as a thick horizontal band of texture surrounded by a background that is dark on the left but gets gradually lighter to the right. The horizontal band itself appears obviously lighter on the left than on the right. Yet in fact the texture is of uniform brightness over its entire width, as you can easily see if you screen off the background with your hands.

Our visual system can deceive us in more subtle ways. Look at Figure 2. This is known as a Kanizsa triangle, after the Italian psychologist

Fig. 1. How uniform is the shading of the horizontal band?

Fig. 2. Do you see a
white triangle?

Gaetano Kanizsa, who worked in Trieste.[1] You will probably see a large
upright white triangle that appears to stand in front of three black circular discs.* This white triangle may appear to you as slightly whiter
than the white of the rest of the figure.

Contours such as those of the illusory white triangle are often called
"illusory contours," since there is not real contour there, as you can
easily see by covering the rest of the figure with your hands while leaving exposed a short section of the "contour." The paper left visible
then appears to be uniformly white, without any contour.

My second general remark is that *any one aspect of the visual information provided by your eyes is usually ambiguous.* It is not enough, by
itself, to enable you to interpret it unambiguously in terms of objects
in the real world before you. In fact, there are often many different
conceivable interpretations.

An obvious example is seeing in three dimensions (3D). If you keep
your head completely still and close one eye you can, to some extent,
still see the world in depth, yet the only visual information we have
comes from the two-dimensional (2D) image falling on the retina of
the open eye. Suppose the object directly before you is a square wire

*The actual shape of one of the black areas—a sectored disc—is usually called a
"pacman."

frame at a certain distance, viewed against a uniform white background (see Fig. 3a). You certainly will see it as a square.

However, the wire might not, in reality, form a square at all, but consist of an *inclined* rectangle with a peculiar shape (see Fig. 3b) so that its projection onto the retina was exactly the same as that of a square frame facing directly toward you. Moreover there are a large number of distorted wire frames that could give the same image on the retina.

This example may appear rather artificial since one seldom looks at the world with one eye closed and the head held still. But suppose you are looking at a photograph or a realistic painting of some scene. Moving your head or using both eyes merely tells you that, in reality, the photograph or picture is flat. Yet in most cases you still see what is represented in the picture in three dimensions.

Certain simple drawings can have several equally plausible interpretations. Consider Figure 4. This consists of twelve black continuous straight lines on the surface of the paper, yet almost everyone sees it as the outline of a 3D (three-dimensional) cube.

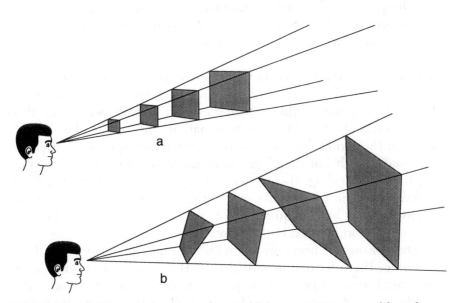

Fig. 3. All the objects—the rectangles—in these two pictures would produce the same pattern on the retina of one eye. The objects in *a* are all the same shape, but differ in size. Those in *b* vary in shape. All the objects have their corresponding corners along the same lines of sight, as shown by the long lines converging near the observer's eye.

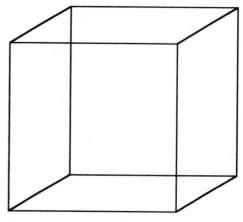

Fig. 4. Keep staring at this cube.
Does its *appearance* change?

This particular drawing—called a Necker cube—has an interesting property. Look at it fairly steadily for a while, and the cube will invert, as if it were being viewed from another angle. After a time the percept switches back to the original one, and so on. In this case there are two equally plausible 3D interpretations of the image, and the brain is uncertain which it prefers. Notice that it only chooses one at a time, not some odd mixture of both of them.

The problems of interpreting the different aspects of the visual image are examples of what mathematicians describe as "ill-posed problems." To any one of them there are many possible solutions, all of which, without any additional information, are equally plausible. To arrive at the veridical solution—the one that corresponds most closely to "what is really out there" (as measured by other tests, such as walking there and feeling it)—we need to apply what, mathematically, are called "constraints." In other words, the system must have been provided with, or have acquired, built-in assumptions about how the incoming information is best interpreted.

The reason you normally see without ambiguity is that the brain combines the information provided by the many distinct features of the visual scene (aspects of shape, color, movement, etc.) and settles on the most plausible interpretation of all these various visual clues taken together.

My third general remark is that *seeing is a constructive process*, meaning that the brain does not passively record the incoming visual information. It actively seeks to interpret it, as the above examples have shown. Another striking example is the process of "filling in." One type of filling in concerns the blind spot, which occurs because the nerves connecting the eye to the brain have to leave the eye at some point, and there is no room for photoreceptors in this little area of the retina (see Fig. 38). Close or cover one eye and gaze straight ahead. Hold up one finger vertically, about a foot or so away from your nose, so that its tip is roughly at the same level as the center of your eye. Move the tip horizontally so that it is about 15 degrees away from the center of gaze, in the direction away from your nose. With a little searching you can find a place where your fingertip becomes invisible. (Be sure to keep your gaze straight ahead.) You are blind for that small area of the visual field.

You are blind there, but no obvious hole appears in your visual field. As I write I am looking out the window of my study at home onto a grass lawn. Even if I close one eye and look straight ahead, I see no hole in my percept of the grass. Astonishing as it may seem, the brain tries to fill in that blind spot with its best guess as to what might be there. Just how it constructs this guess is what psychologists and neuroscientists are trying to find out. (I discuss the filling-in process more fully in Chapter 4.)

At the beginning of this chapter I have put the phrase "Seeing is believing." In normal parlance this means that if you see something you can believe it is really there. I would stress a quite different interpretation of this cryptic phrase: What you see is not what is *really* there; it is what your brain *believes* is there. In many cases this will indeed correspond well with characteristics of the visual world before you, but in some cases your "beliefs" may turn out to be wrong. Seeing is an active, constructive process. Your brain makes the best interpretation it can according to its previous experience and the limited and ambiguous information provided by your eyes. Evolution has seen to it that your brain usually does this with remarkable success, but not always. Psychologists are interested in visual illusions because these partial failures of the visual system can give useful clues about the way the system is organized.

How then should we think about vision? Let us start with the naive views of someone who has not worried about the problem. Clearly I appear to have a "picture," in my head, of the visual world in front of me. Yet few people believe that there is an actual screen, somewhere inside their brain, that produces patterns of light corresponding to the visual world outside the head. We all know that there are machines that can do this, such as television sets, yet on opening a person's head we do not find brain cells in an ordered array, emitting lights of various colors. Of course, the information in the TV picture need not be embodied only on its screen. If you were to produce computer art, using a special computer program, you would find that the information needed to form your picture on the screen is not stored in the computer as patterns of light. Instead, it is stored in the computer's memory as a series of electrical charges in its memory chips. It may be stored there as a regular array of numbers, each number representing the light intensity at some point. Such a memory does not *look* like a picture, but the computer can use it to produce the picture on its screen.

Here we have an example of a *symbol*. The information in the computer's memory is not the picture; it symbolizes the picture. A symbol is something that stands for something else, just as a word does. The word *dog* stands for a particular sort of animal. Nobody would mistake the word itself for the actual animal. A symbol need not be a word. A red traffic light symbolizes "stop." Clearly, what we expect to find in the brain is a representation of the visual scene in some symbolic form.

Well, you might say, why should there not be a *symbolic* screen in the brain. Suppose the screen were made of an ordered array of nerve cells. Each nerve cell would handle the activity at one particular "point" in the picture. The activity of the cell would be proportional to the intensity of the light at that point. If there were a lot of light there, that nerve cell would be very active; if no light, then it would be inactive. (By having a set of three nerve cells for each point we could deal with color as well.) Thus the representation would be symbolic. The cells of this postulated screen do not produce light, but some form of electrical activity that symbolizes light. Why should this not be all we need?

The trouble with such an arrangement is that it would not be "perceiving" anything except little individual patches of light. It could not *see*, any more than your television set can see. You can tell a friend: "Let me know when that nice young woman starts reading the news," but it is no use trying to wire up your television set to do this. It has no way built into it for recognizing a woman, let alone a particular one performing a particular action. Yet your brain (or your friend's brain) can do this with little or no apparent effort.

So the brain cannot get by with just sets of cells that merely show what sort of light intensity is where. It must produce a *symbolic description at a higher level*, probably at a series of higher levels. As we have seen, this is not a straightforward matter, since it must find the best interpretation of the visual signals given its past experience. Thus, what the brain has to build up is a many-levelled interpretation of the visual scene, usually in terms of objects and events and their meaning to us. As an object, like a face, is often made up of parts (such as eyes, nose, mouth, etc.) and those parts of subparts, so this symbolic interpretation is likely to occur at several levels.

Of course, these higher-level interpretations are *implicit* in the pattern of light falling on the retina, but this is not enough. The brain must make those interpretations *explicit*. An explicit representation of something is what is symbolized there without further extensive processing. An implicit representation contains that information but much further processing is needed to make it explicit. It would be easy to add a simple gadget to a TV set so that it would signal whether a particular point on its screen had a red dot there. To wire it up so that it flashed a light when it saw the woman's face anywhere on the screen would require much further processing, so much more that today we cannot yet build the very complicated apparatus needed to do it.

When something has been symbolized explicitly, this information can easily be made available so that it can be used, either for further processing or for action. In neural terms, "explicit" probably means that nerve cells must be *firing* in a way that symbolizes such information fairly directly. Thus it is plausible that we need an *explicit multilevel, symbolic* interpretation* of the visual scene in order to "see" it.

It is difficult for many people to accept that what they see is a symbolic interpretation of the world—it all seems so like "the real thing." But in fact we have no direct knowledge of objects in the world. This is an illusion produced by the very efficiency of the system since, as we have seen, our interpretations can occasionally be wrong. Instead, people often prefer to believe that there is a disembodied soul that, in some utterly mysterious way, does the actual seeing, helped by the elaborate apparatus of the brain. Such people are called "dualists"— they believe that matter is one thing and mind is something completely different. Our Astonishing Hypothesis says, on the contrary, that this is not the case, that it's all done by nerve cells. What we are considering is how to decide between these two views experimentally.

---

*The use of the word *symbol* should not be taken to imply the literal existence of a homunculus. It merely means that the firing of the neuron (or neurons) is highly correlated with some particular aspect of the visual world. Whether such a symbol should be thought of as a vector (rather than just a scalar) is a tricky question I shall not consider here. In other words, how distributed is a single symbol?

4

The Psychology of Vision

"When we trace the history of psychology we are led into a labyrinth of fanciful opinions, contradictions, and absurdities intermixed with some truths."

—Thomas Reid

I hope I have convinced you that seeing is not as simple as you might have thought. It is a constructive process in which the brain responds in parallel to many different "features" of the visual scene and attempts to combine them into meaningful wholes, using its past experience as a guide. Seeing involves active processes in your brain that lead to an explicit multilevel, symbolic interpretation of the visual scene.

Let us now look at some of the basic operations that the brain must perform in order for us to see objects, their position (relative to us and to one another), and certain of their attributes, such as shape, color, motion, and so on. Perhaps the most important thing to realize is that, in the visual field, *objects are not given to you as such*. Each object is not clearly and unambiguously marked for you. Your brain has to use various clues to enable it to group together those parts of the visual scene that correspond to a single object. In the real world this is often not easy. The object may be partly hidden or seen against a confusing background.

An example may make this clearer. Look at the photograph in

Fig. 5. One face,
or parts of four faces?

Figure 5. You will see immediately, without any apparent effort, that it represents the face of a young woman looking out of a window. But look closely at it. The wooden slats of the window frame cut up the woman's face into separate pieces. Yet you do not see four separate slices of four separate faces that, accidentally, have come to be near each other. Your brain groups the four parts together and interprets them as a single object—a face—partially occluded by the slats in front of it. How is this grouping performed?

This was one of the major interests of the Gestalt psychologists Max Wertheimer, Wolfgang Köhler and Kurt Koffka. The movement started in Germany around 1912, although when the Nazis came to power all three left Germany, ending up in the United States. My dictionary defines "gestalt" as "an organized whole in which each individual part affects every other, the whole being more than the sum of its parts."* In other words, your brain must actively build up these

---

*As I have explained in Chapter 1, this is certainly true if "sum" is used in too naive a manner.

"wholes" by finding which combination of the parts seems the most likely to correspond to the relevant aspects of the object in the real world, basing its estimates on your previous experience and on the experience of your distant ancestors, which is embedded in your genes.

Obviously, what is important is the *interaction* of the parts. The Gestaltists attempted to classify the types of interaction that appeared to be common in the visual system, calling them laws of perception.[1] Their laws of grouping included proximity, similarity, good continuation, and closure. Let us look at each of these in turn.

Their Law of Proximity stated that we tend to group together things that are close to one another and more distant from other (similar) objects. This is obvious in Figure 6. This consists of many small black dots arranged in a regular rectangular array. Your brain could group the dots together in either vertical lines or horizontal lines. The fact that you see them in vertical lines is because the distance from one dot to its nearest neighbors is shorter in the vertical direction than in the

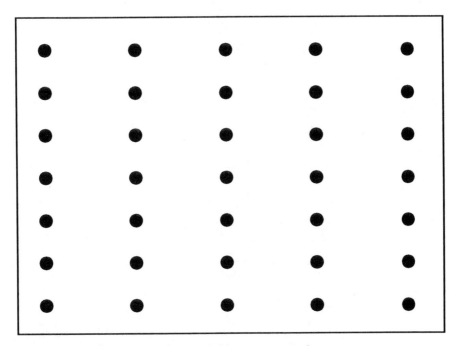

Fig. 6. Do the dots suggest horizontal lines or vertical ones?

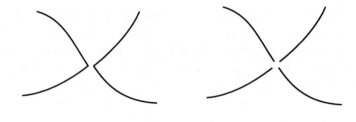

Fig. 7. Do you
see the bottom
figure as two
lines crossing
each other?

a.

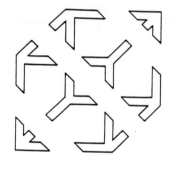

b.

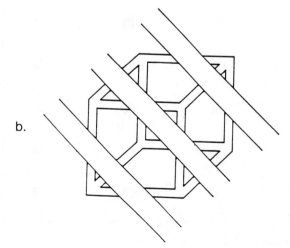

Fig. 8. What objects
do you see?

horizontal one. Other experiments show that proximity usually means "proximity in space" rather than proximity on the retina.

The Gestaltists' Law of Similarity means that we group things together if they have some obvious visual property in common, such as color or direction of movement. If you see a cat moving you group the parts of its body together because, on the average, all the parts move in the same direction. A cat creeping through the undergrowth can be recognized for the same reason. If it stays perfectly still it may be difficult to spot that it's there.

Their Law of Good Continuation is illustrated in Figure 7, which shows two curved lines crossing each other. We do indeed see this as two lines, rather than as four lines, all meeting at a point, or two V-shaped figures meeting, as shown in the upper part of the figure. We also tend to see interrupted lines as continuous lines with something occluding part of them.

Look at the strange set of eight objects in Figure 8a. The two in the middle look like the letter Y. The other six resemble distorted arrows. Now look at Figure 8b. You will probably see a 3D cubic frame occluded by three diagonal slats. The strange objects are parts of both figures, but you see the cube more easily in the second one because it appears to be a single object occluded by the slats. The first figure, lacking any strong suggestion of occlusion, looks more like eight independent objects.

Closure is most easily seen in line drawings. If a line forms a closed, or almost closed, figure we tend to see a surface figure bounded by a line, rather than just a line.*

The Gestaltists also had a general principle they called Prägnanz—roughly translated as "goodness." The basic idea here is that the visual system arrives at the simplest, most regular and symmetrical interpretation of the incoming visual information. How does the brain decide what is the "simplest" interpretation? A modern view is that the best interpretation is the one that needs only a little information (in the technical sense) to describe it, whereas a bad interpretation needs more.†

*More recently, Stephen Palmer, a psychologist at the University of California, Berkeley, has proposed[1] two further laws: common region and connectedness. Common region, or enclosure, means grouping within the same perceived region. Connectedness refers to the powerful tendency for the visual system to perceive any uniform, connected region as a single unit.

†This may depend somewhat on what "primitives" one takes as given in estimating the information content.

Fig. 9. Can you see the dog?

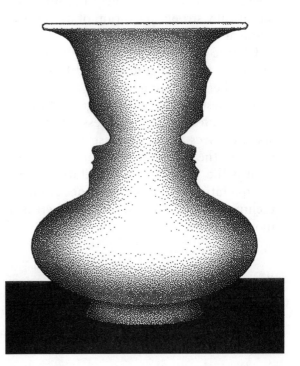

Fig. 10. A vase,
or two profiles?

Put another way, the brain usually prefers a sensible interpretation to a freak one, meaning that the interpretation would not be radically altered by a small change of viewpoint. This may be so because, in the past, while looking at an object, you were often moving through the visual world, so your brain recorded different aspects of that object as belonging to one thing.[2]

The Gestaltists' Laws of Perception should not be regarded as rigid laws but as useful heuristics. As such they can serve as a convenient introduction to the problems of vision. Exactly what processes are really operating to produce the appearance of these "laws" is what many visual psychologists are trying to discover.

An important operation in vision, as the Gestaltists recognized, is to separate figure from ground. The object to be recognized is called the "figure"; its surroundings are the "ground." This separation may not always be straightforward. Look carefully at Figure 9. If you have never seen this before you may have difficulty seeing any recognizable object there. After a while you will probably realize that part of the picture represents a Dalmatian dog. The separation of figure from ground in this case has deliberately been made more difficult.

It is possible to construct images in which figure-ground separation is ambiguous. Look at Figure 10. At first glance it looks like a vase. Keep looking at it and you may see instead the profiles of two faces. Whereas the vase was originally the figure, the profiles now make the figure, and what was the vase has become the ground. It is difficult to see both interpretations at exactly the same moment.

The brain depends upon many strong hints to decide what visual features belong to a single object, broadly along the Gestalt Laws of Perception just described. Thus, if a certain object is fairly compact (proximity), shows us a well-defined outline (closure), is moving in one direction (common fate), and is all the same color (similarity), we are likely to recognize that it's a moving red ball.

It is important that an animal can do this well, otherwise it might not easily spot a predator or a prey or other forms of food, such as an apple. It must be able to segregate the figure from the ground. What we call camouflage is an attempt to deceive these processes. Thus, camouflage breaks up the continuity of the surface (as in battle dress) and produces confusing outlines to disguise the true outline. Colors are used that are likely to blend into the background. A stalking cat moves cautiously and freezes from time to time to avoid giving motion clues to its prey. It has even been suggested that our good

Fig. 11. Just lines.

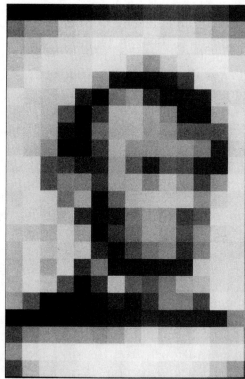

Fig. 12. Try looking at this
from a distance.

color vision evolved to enable our primate ancestors to spot red fruit against a confusing background of green leaves. What gives us so much visual pleasure may be, in origin, a device to spot our food and to break camouflage.

We know something of the earliest stages of visual processing, partly from studies on the eye and the brain (see Chapter 10). Almost the first operation to be performed is to remove redundant information. The photoreceptors in the eye respond to the intensity of the light falling on them. Suppose you are looking at a fairly uniform, smooth, white wall. Then a whole patch of photoreceptors in your eye will be responding to the light in much the same way. Why send *all* this information to the brain? It is better for the retina, at the back of your eye, to process the visual information, so that the brain is told where the light intensity changes in space, at the edges of the wall. If no spatial change is present, across some region of the retina, no signal is sent. The brain then has to infer that "no signal" means "no change" and that part of the wall is visually uniform.

To some extent, as we shall see in later chapters, the brain processes different sorts of visual information in somewhat separate parallel streams. Thus, it makes sense to study separately how we see shape, motion, color, and so on, even though these processes interact somewhat.

Let us start with shape. It is obviously useful for the brain to extract outlines. This is why we respond so easily to line drawings. You can interpret a line drawing of a visual scene even if it has no shading, texture, or color in it (see Fig. 11). It turns out that different elements in the brain respond better to fine detail, others to less fine detail, and still others to coarser spatial changes. If you saw only the latter, the world would look blurred and out of focus compared to your usual view of it. Psychologists speak of "spatial frequency," a high spatial frequency corresponding to fine detail. Low spatial frequency corresponds to more gradual spatial changes of the image.

Look at Figure 12. You can probably see it as a composite of small squares, each a uniform shade of gray. Now make the picture blurred (by removing your glasses, by squinting at the picture through half-closed eyes, or by putting the illustration on the far side of the room) and you will probably recognize the face of Abraham Lincoln. The fine detail in the picture—the edges of the squares—was interfering with the process of recognition. With blurred vision these edges become less obvious. You can then recognize the face, even though it appears somewhat fuzzy, using only the lower spatial frequencies in the image. Of course, normally, both low and high spatial frequencies help one to interpret the image.

One of the most difficult problems the brain has to face is how to extract depth information from a 2D image. It needs this not only to discover how far away things are from the observer but also to see the 3D shape of each object. It helps to have two eyes, but the shape of an object can often be seen using only one eye, or by looking at a photograph of it. What clues does the brain use to see depth in a 2D picture? One clue is the shading of an object, produced by the angle of the incident light. Look at Figure 13. You will probably see it as a row of four hollows and a second row of bulges in an otherwise flat surface. This impression of depth comes from the shading produced by the incident light.

Incidentally, this interpretation is ambiguous. Stare at it for some time, or turn the page upside down, and you will probably now see the hollows as bulges and vice versa. (Notice that they all change together.) Your brain initially assumes that the light comes from one side. If

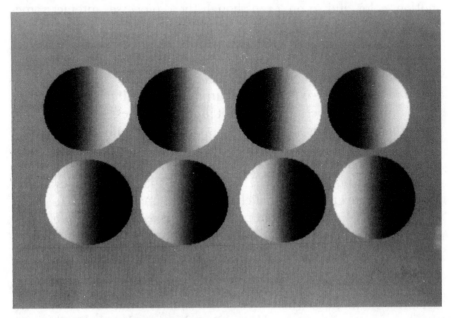

Fig. 13. Hollows or bulges? Keep looking.

the light really comes from the other side, then the same shading corresponds to a different shape, as you have just seen.

Another very compelling clue is "structure from motion." This means that if the shape of a stationary object is difficult to see (often because some of the clues to its 3D shape may be missing), it may be easier to recognize if the object is rotated somewhat. A picture of a complicated molecular model, made of balls and spokes, is not easy to grasp when projected onto a screen during a lecture. Yet if a movie of the model is shown, with the model rotating, the 3D shape springs into view. You may have seen this toward the end of the TV program "Life Story," where the model of DNA is made to rotate to celestial music.

To see in 3D, it is not enough to see each object in 3D. You need to see the whole scene in 3D so that you know which objects are near to you and which farther away. There are two strong clues to this even in 2D pictures.

The first is perspective. This is vividly demonstrated by a distorted room known as an Ames room after Adelbert Ames, who invented it. This room is viewed from outside, through a small hole, using only one eye to avoid any clues from stereopsis. The room appears to be rectangular, but it has actually been built with one side stretched so that one corner is much higher and also farther from the observer than it would be if the room were square (see Fig. 14). When I looked into such a room at the Exploratorium in San Francisco I saw some chil-

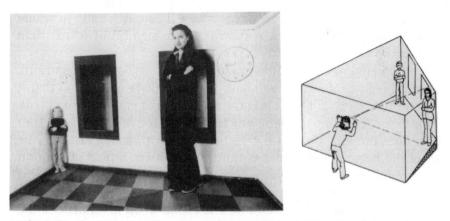

Fig. 14. On the left, what the Ames room might look like viewed through a peephole; on the right, a schematic diagram of the room and the observer.

dren running about in it. On one side they appeared tall (because, in reality, they were quite near me) whereas on the other side the same child appeared short (being now farther away). As they ran from side to side (actually from the near corner to the farther corner and back again) they appeared to change size quite dramatically. I know full well that children cannot alter their size in this way, yet the illusion was so compelling that I could not immediately shake it off. The apparent size of each child was set by the (false) perspective given by the walls of the room. This illusion, like many others, is not easily corrected by "top-down" effects—what the highest levels of the brain know about the basis of the illusion.

The other strong clue is occlusion—when an object nearer to you partly hides one farther away. We have already seen this in the picture of the girl's face behind the slats of a windowpane (Fig. 5). To use this clue, the brain has to deduce that the different parts of the occluded object belong together, as discussed at the beginning of this chapter.

Lines can produce two surprising effects related to occlusion. You saw the first type in the Kanizsa triangle (Fig. 2). The (illusory) boundary of the white triangle is produced by the continuation of the straight edges of the pacmen. The other type is shown in Figure 15.

In this case the illusory boundary is mainly due to the lining up of the *ends* of a set of lines (called "line terminators"). A "line" in the visual field can occur for many reasons—a pattern on an object, such as a shirt, or the stripes on a zebra—or a shadow and so on. An object occluding a background will often interrupt lines in that background. In such cases, the illusory contours produced by line terminators will outline the object, as in the artificial case shown in Figure 15. As the psychologist V. S. Ramachandran has said, "There is a sense in which illusory contours are more real (that is, more significant to us) than real contours."

Another clue to distance is a gradient of texture. Look at Figure 16. You see a picture of a lawn, and have the immediate impression that it is receding from you. This is because the blades of grass become smaller, *on the paper*, in a systematic way toward the top of the picture. Your brain does not see this as a flat vertical wall, with large grass growing out of the bottom and small grass at the top but as a field of uniform grass stretching away from you.

There are other clues to depth. One is apparent size—a familiar object produces a smaller image on the retina when it is farther away, so if its apparent size is small this may suggest to the brain that the object is farther away. Another is that landscapes usually look bluer at

Fig. 15. Do you see an unfamiliar, white shape?

Fig. 16. Do you see this as receding?

a distance. All these clues have been exploited by artists, especially since the discovery of perspective in the Renaissance. Good examples are the paintings of views of Venice by Canaletto.

Now let us turn to a major source of depth information.*This is called "stereopsis," and it depends on the fact that each of our two eyes has a slightly different view of the world. It was the physicist Sir Charles Wheatstone who, in the middle of the nineteenth century, first clearly demonstrated that two such views, properly presented, can give a vivid impression of depth. (Wheatstone is also remembered because, while waiting to deliver a Friday Evening Discourse at the Royal Institution in London, he became so nervous that he ran away. Since then the Discourse Lecturer, who waits in a small room for a quarter of an hour before giving the lecture, is routinely locked inside the room.) Wheatstone invented a stereo viewer—the one that became popular in late Victorian times had a simpler design—that enabled each eye to look at a separate photograph, each taken from a slightly different position. This difference in position produces two views that are not exactly the same. The brain is able to detect the differences between the two views, referred to technically as "disparities," and as a result the scene in the photograph appears in vivid depth, as if it were directly before you.

You can easily test stereopsis for yourself by closing one eye as you look at the nearer items in the real scene in front of you. For most people the impression of depth is not nearly as strong as it is when using both eyes. (Of course, even with one eye closed you have a fairly good impression of where objects are in depth because of the other clues discussed above.) Another obvious example is a realistic painting or photograph of a building, a city, or a landscape. In this case the two eyes allow the brain to deduce that the surface of the painting is flat. In fact, you will find the impression of depth more vivid if you look at the picture with only one eye while at the same time screening the picture frame from view with your hands and standing in a position where there are no reflections from the glass. These actions remove some of the clues to the flatness of the surface of the picture, with the result that the clues the artist has put into the picture to convey depth have a stronger effect.

Stereopsis works best for objects fairly close to you, since then the disparity between your two views is largest. Obviously, for you to have

---

*A small percentage of humans appear to lack true stereo vision.

a view of the same object with both eyes, the object needs to be more or less in front of you and not so far removed to one side that your nose prevents one eye from seeing it. Predators that have to pounce on their prey, such as cats and dogs, usually have frontally facing eyes. They can use stereopsis in grappling with their prey. For other animals, such as rabbits, it is more useful to have the eyes placed more to the side of the head, so that they can scan for predators over a wider field of view. Their stereopsis is more limited than ours, since their eyes have less visual overlap.*

What about motion? The visual system is very interested in movement, for obvious reasons. When you watch a movie you frequently have a vivid impression of moving objects although what is actually appearing on the screen is a rapid succession of still pictures. This phenomenon is known as "apparent movement." In this rather artificial situation vision may go wrong. The spokes of the wheels of a car or a cart may appear to move in the wrong direction. It is well known, in general terms, why this happens. Roughly speaking, the brain connects a particular spoke in one image with the nearest one in the next image. This may not be the identical spoke but a neighboring one, because of the wheel's rotation. Since all the spokes look fairly alike, the brain may link together two different spokes in the two successive pictures. If the two positions linked together are identical (relative to the car) the wheel appears to be stationary. If the speed of rotation is slowed a little, the spokes of the wheel will appear to move backward. It can often happen, especially in older movies, that as the car slows, the spokes appear to change direction relative to the movement of the car. Psychologists have done many experiments trying to decide exactly what conditions are needed to give good apparent motion.

Another motion effect is the so-called barber's pole illusion. The pole has helical stripes on it. When it is made to rotate about its long axis the stripes appear not to rotate but to move along the length of the pole, usually upwards. (This is discussed more fully in Chapter

---

*There are considerable theoretic problems about how the brain uses this disparity. For example, it has to know which feature of the visual image in one eye corresponds with which feature in the other eye. This is called the "correspondence problem." It was originally thought that the brain had first to recognize the object to solve this problem. Some brilliant experiments by the Hungarian psychologist Bela Julesz, then at Bell Laboratories, using stereograms made of random dots, showed that the "correspondence" between the two images could be established at a low level in the visual processing, *before* the object was recognized.

11.) Thus our perception of motion is not always completely straight-forward. In this case you do not see the *local* motion of each bit of the stripes; instead, your brain mistakenly imagines the global motion of the whole pattern.

The perception of motion by the brain is handled by two main process-es, called (somewhat inaccurately) the "short-range system" and the "long-range system." The former is believed to occur at an earlier stage of processing than the latter. The short-range system does not recognize objects but merely the changes in the patterns of light sensed by the retina and conveyed to the brain. It extracts movement as a "primitive" without knowing what is moving. In other words, this simple aspect of motion can usefully be regarded as a primary sensation. It operates automatically—that is, it is not influenced by attention.

It is suspected that the short-range system can segregate figure from ground* using movement information and that it is responsible for the motion after-effect, sometimes called the "waterfall effect." (If you gaze at a waterfall for some time and then shift your gaze to the adjacent rocks, they will briefly appear to move upwards.) There is now some doubt about this; it was recently shown[3] that the motion after-effect can be influenced by attention.

The long-range motion system appears to register the movement of objects. Instead of just registering movement as such, it registers *what* is moving from one place to another. This can be influenced by attention.

As an (oversimplified) example, consider a situation in which a red square is briefly flashed onto a screen followed, after a certain interval, by a blue triangle flashed a little distance away. If the parameters (of distance and time) are chosen so that the long-range system is pre-dominant, then the observer sees the apparent motion of a red square changing into a blue triangle as the object moves from one position to the other. If, on the other hand, the chosen parameters mainly acti-vate the short-range mechanism (because the distance and the time interval are both short), then the observer will see movement but he may not see the moving object. He senses motion without knowing what is moving. In most circumstances, both systems will be in opera-tion to some extent. Only very carefully designed stimuli can activate just one of them.

*   *   *

*This segregation of figure from ground poses a difficult theoretical problem, since the brain must segregate figure from ground without knowing what the figure is.

The brain uses motion clues to derive additional information about the changing visual environment. I have already described how, in some cases, it can derive structure from motion, but it can use motion in other ways. An object coming directly toward your eyes has an expanding image on your retina. If an object on a screen is suddenly made to increase in size, you have the impression that it is rushing toward you (even though the screen is still the same distance away). This kind of movement of the visual image is called "dilation." It produces such a vivid effect that one would suspect that there is a special part of the brain that responds to dilation of the image, and indeed this has been found (see Chapter 11).

Another use for the visual motion system is to guide the way you move through the world. As you walk forward, with your eyes looking ahead, the visual scene slips past you on either side and above and below your head. This movement of images on the retina is called "visual flow" and is a great help to a pilot in landing a plane. A one-eyed pilot (who therefore lacks stereopsis) can land a plane quite well guided by the information from visual flow. Where there appears to be no flow is the point toward which he is moving. All around this point the visual objects appear to move away from that point, although at varying rates (see Fig. 17). This visual information helps him pilot the plane toward the correct point on the landing strip.

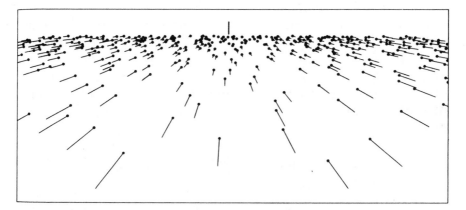

Fig. 17. Landing a plane using only one eye. The observer is supposed to be moving toward the spot on the horizon marked by the vertical line. A series of black dots has been scattered about the visual field. The thin line attached to each dot shows the direction the dot would appear to move due to the observer's forward motion. The length of each line is proportional to the dot's speed of movement.

The perception of color is also not as straightforward as it seems. The basic account relates it to the various types of photoreceptors in the eye. Any one type of these responds to photons (the particles of light) of only a limited range of wavelengths. It is important to realize that *how* a single photoreceptor responds does not depend upon the wavelength of the incoming photon. It either captures the photon or it doesn't. If it does capture it, the effect is the same whatever the wavelength of the photon. However, the *probability* of it responding does depend on the photon's wavelength. Some wavelengths have a bigger chance of exciting it, others less so. For example, it may respond very often to "red" photons, but only rarely to "green" ones.

An average rate of response to a stream of incoming photons may be due either to a few photons of a favored wavelength or to many photons of a less-favored wavelength; the receptor cannot tell which. At first reading, this may all sound rather complicated but the take-home lesson is that if an eye has only one type of photoreceptor, your brain lacks any information about the wavelength of the incoming light and you can see only in black and white. This happens in very dim light when the photoreceptors called "cones" are not active; only those called "rods" are. These are of only one type and thus all respond to wavelength in the same way. This is why if you look at flowers in the garden at night, in a very dim light, they often appear colorless.

To obtain color information, more than one type of photoreceptor is needed, each with a *different* wavelength-response curve. These curves overlap somewhat, but since a stream of incoming photons, all of one wavelength, will on average excite the different photoreceptors by different amounts, the ratios of their excitations can be used by the brain to decide the "color" of the incoming light at that point on the retina.

It is well known that most humans have three kinds of cones (crudely, a short wavelength, a medium wavelength, and a long wavelength: often called, inaccurately, blue, green, and red, respectively), although a small percentage of men lack the "red" cones and are thus partially colorblind.* They may have trouble distinguishing between red and green traffic lights.

*Strictly speaking, we are all colorblind—that is, apart from wavelengths, such as those in the ultraviolet, which we cannot see at all, it is possible to construct any number of wavelength distributions that, while somewhat different if measured by a suitable physical instrument, look exactly the same to us. This is because each of these equivalent distributions has been designed to excite the three cones in exactly the same ratios. With minor reservations, our response to any distribution of wavelengths can be matched by a suitable combination of just three wavelengths, a fact established in the nineteenth century. In mathematical jargon, color is three-dimensional.

*   *   *

This is the basic explanation of how we see color, but it needs modification in a number of ways, of which I will mention only one. This is now known as the Land effect after Edwin Land, the inventor of Polaroid, who studied it extensively. What Land showed in a dramatic manner was that the color of a patch in the visual field does not depend solely on the wavelengths entering the eye from that patch but also on the wavelengths entering from other regions of the visual field.

Why should this be so? What enters the eye depends not only on the reflectance properties (the color) of the surface in question but also on the wavelengths of the light falling on that surface. For this reason, a woman's colored dress should look very different when seen in daylight compared to its colors seen under candlelight. The brain, therefore, is not primarily interested in the combination of reflectance and illumination but in the color properties of the surfaces of objects. It tries to extract this information by comparing the eyes' response to several regions in the visual field. It does this by using the constraint (the assumption) that the color of the illumination is much the same everywhere in that scene at that particular time, even though it may be considerably different on other occasions. If the illumination is pink, this makes everything somewhat pink, and the brain then attempts to compensate for this. That is why a red fabric in daylight still looks fairly red in artificial light although, as we all know, it does not look exactly the same since the compensation mechanism does not work perfectly.

I shall mention only in passing certain other visual constancies. An object looks roughly the same even if we do not look at it directly, so that it falls on a different part of the retina. We recognize it as the same object even when we see it at another distance, so that the size of its image on the retina is larger or smaller, or even if it is rotated somewhat. We take these different constancies somewhat for granted, but a simple vision machine would not be able to perform such feats unless it had built-in devices for doing so, as the developed brain must have. Exactly how the brain does all this is still somewhat uncertain.

Movement and color have a strange relationship. The short-range movement system in the brain is somewhat colorblind and sees mainly in black and white. This is shown most easily by projecting onto a screen a moving pattern made of only two uniformly bright colors, say red and green.

The relative brightness of the two colors is then adjusted so that they appear to the observer to be equally bright. This has to be done for each individual, since my balance point may not be exactly the same as yours.* Such a balanced condition is referred to as "isoluminance."

If now you looked at a pattern on a screen of moving red objects on a green background, with the two colors adjusted to be isoluminant, the movement appears much slower than it really is and may even stop altogether. (This is especially true if you direct your gaze to one side of the screen.) The reason is that the black-and-white system in your brain sees the screen as a uniform gray (since the two colors have been made equally bright) and so the short-range motion system registers little or no movement.

All these examples show that the brain can extract useful information from many somewhat different aspects of the visual scene. What does it do if the information provided is incomplete? A good example is provided by the blind spot. As described in Chapter 3 (page 31), you have a blind spot in each eye that the brain "fills in" so that even with one eye closed you do not see a hole at that point in the visual scene. The philosopher Dan Dennett does not believe that there is a filling-in process. In his book *Consciousness Explained,* he has argued, correctly, that "an absence of information is not the same as information about an absence." He continues: "In order for you to see a hole, something in your brain would have to respond to a contrast: *either* between the inside and outside edge—and your brain has no machinery for doing that at this location—*or* between before and after." Therefore, he argues, there is no filling-in. Just the absence of any information that there is a hole there.

This argument, however, is flawed, since he has not proved that information is *not* inferred in the region of the blind spot. His argument is only that the brain *may* not make the inference. Nor is it correct to say that the brain definitely does not have the necessary machinery. A careful study of the brain shows that it does have nerve cells there that might do the job (see Chapter 11).

The visual psychologist V. S. Ramachandran, who works in La Jolla in the Psychology Department of the University of California, San Diego, has done a neat experiment[4] to refute Dennett. (Everyone likes to show that philosophers are wrong.) He showed a subject a picture

---

*Even for a single observer, the balance point may be slightly different for objects in the line of gaze compared to objects in the visual periphery.

of a yellow annulus (i.e., a thick ring, or doughnut—see Fig. 18b). The subject had to keep his eyes still and view the world with only one eye. Ramachandran positioned the yellow ring in the subject's visual field so that its outer rim was *outside* the subject's blind spot (for his open eye), while its inner rim was *inside* it (see Fig. 18b). The subject reported that what he saw was not a yellow ring but a complete homogeneous yellow *disk* (Fig. 18c). His brain had filled in the blind region, thus turning a thick ring into a uniform disk.

To underline this result, Ramachandran placed several other similar rings in the subject's visual field. When this display was switched on (with one ring around the blind spot, and the others elsewhere), the

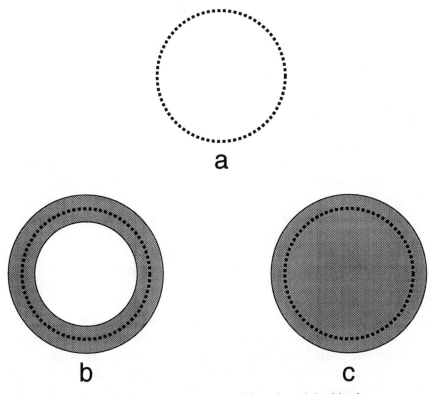

Fig. 18. *a:* A schematic representation of the edge of the blind spot, shown dotted. *b:* The shaded doughnut represents the (yellow) annulus presented to the open eye. It was positioned to lie over the margin of the blind spot. *c:* This represents what the subject saw—not a yellow annulus but a complete yellow disk—due to the filling-in process. Note that he never sees the outline of the blind spot, shown here as a dotted circle.

subject reported not only that he saw a complete disk in the region of his blind spot but also that it "popped out" at him, meaning that his attention was immediately drawn to it. This is exactly what would happen to you, with both eyes open, if you were shown a random array of yellow rings plus one solid disk. The disk, being so obviously different, would pop out at you. As Ramachandran had said, this suggests that you really do fill in the blind spot—you don't just ignore what is there. For how can something you ignore actually pop out at you?

What we see in the blind spot is not easy to study since it is 15 degrees away from the center of gaze and, as I described above, we do not see very clearly there. Ramachandran and the English psychologist Richard Gregory have done experiments[5] on what they call an "artificial scotoma" that can be produced much closer to the center of gaze. (Dennett mentions this work in a footnote but is unhappy with their results.) More strikingly, Ramachandran and his colleagues[6] have now examined patients with a small, damaged region not in their eyes but in a visual part of their brain. Such a patient cannot truly see what is in the corresponding portion of his visual field—he has a blind patch there—but there is no doubt that, given time, his brain fills it in with plausible guesses from the surroundings.

Schematic illustrations of some of their results are shown in Figure 19. Two vertical lines on a cathode ray screen were placed in line, one above and one below the blind patch; after a few seconds the patients saw the line completed across the gap. One patient reported that when the lines on the screen were switched off he "saw a very vivid white 'phantom' of the filled-in portion of the line" persisting for several seconds! Even more astonishing, when the two patients were shown two *misaligned* vertical lines (see Fig. 19c) they at first saw them as misaligned, but then the lines started moving or "drifting" toward each other and eventually became collinear. The brain then filled in the gap to make the two lines appear as a continuous one (see Fig. 19d). This horizontal movement of the lines (remember, the lines in the real world were completely stationary) was reported to be very vivid and both patients seemed very surprised and intrigued by the phenomenon.

In other experiments it was shown that not every aspect of vision filled in at the same time. Shape, motion, texture, and color could fill in at different times. For example, if the visual field consisted of a lot of moving, random red dots, for one patient the color started "bleeding" into the blind patch almost immediately, followed five seconds later by the dynamic pattern of moving dots.

Notice that there is an important difference between these results

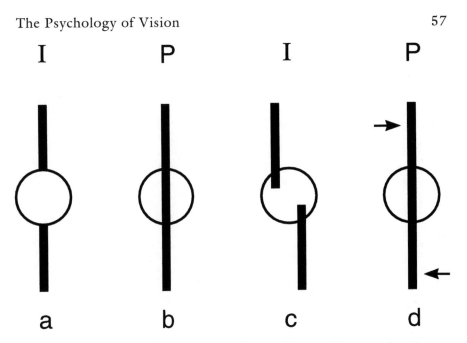

Fig. 19. The figures marked *I* (image) indicate what was shown to the subject; those marked *P* (percept) show what he eventually saw. The circle represents diagrammatically the scotoma—the small area in the visual field corresponding to the brain region that had been inactivated in the subject's brain, in the primary visual cortex on one side of the head.

for a blind patch in the brain and those for the true blind spot in the eye. In the latter case the filling-in is almost immediate. In the case of brain damage the process take several seconds, probably because the damage has removed the apparatus in the brain that fills in rapidly.

Filling-in is probably not a special process peculiar to the blind spot. It is more likely that, in one form or another, it occurs at many levels in the normal brain. It allows the brain to guess a complete picture from only partial information—a very useful ability.

We've now had a few glimpses of the complexity of visual psychology. Obviously, seeing is not a simple matter. It is very different from what you might have guessed from everyday experience. How it all works is not yet fully understood and involves many experiments and concepts that I have had to omit. The next chapter will broaden our approach by touching on two other aspects of seeing—attention and very short term memory—that are closely related to visual awareness, and will introduce the difficult subject of the time taken by different aspects of visual processing.

5

Attention and Memory

"You're not paying Attention," said the Hatter. "If you don't pay him, you know, he won't perform."
                                        —after Lewis Carroll

Everyone knows the common meaning of the phrase "you're not paying attention." This may be because your attention was distracted, because you are drowsy, or for some other reason. Psychology distinguishes between "arousal" (or alertness) and "attention." Arousal is a general condition affecting all of one's behavior, as you may notice when you first wake up in the morning. Attention implies to psychologists, as William James said, "withdrawal from some things in order to deal effectively with others."

Our main concern is with visual attention rather than the attention needed to listen to music or to perform some action. Recall that attention is thought to assist at least some forms of awareness. One form of visual attention is eye movement (often assisted by head movements). Because we see more clearly close to our center of gaze, we get more information about an object if we direct our eyes in that direction. We get coarser information (at least about shape) from objects we are *not* looking at directly.

What controls eye movements? Such movements range from reflex-like responses, such as those to a sudden movement at some point outside our center of gaze, to willed eye movements ("I wonder what

59

he's doing over there"). All forms of attention are likely to have both reflex and willed components.

An example of selective attention in hearing is a subject concentrating on the sounds coming from earphones into one ear while trying to ignore different sounds coming into the other ear. Many of the sounds coming into the unattended ear do not reach consciousness but it can be shown that they may leave some trace in the brain and may sometimes influence what is heard in the attended ear. At some level they are being registered by the brain.

Attention, then, filters out *unattended* events. An attended event is reacted to more rapidly, at a lower threshold and more accurately. Attention also makes the event somewhat easier to remember. In the past, psychologists did not concern themselves with what was happening inside the skull; they mostly studied attention by measuring how it affected the speed of response, the level of errors, and so on. In other words, they studied the *consequences* of attending to an event (compared to not attending to it) and tried to deduce the likely mechanisms of attention from the pattern of their results.

It is surprising what you cannot do if you hold your eyes still. Suppose a random pattern of dots is flashed onto a screen for a brief period, too short for you to start an eye movement. Can you say how many there are? If there are only three or four dots you can report their number without error, but if there are six or seven or more you will make mistakes. This is not just due to the briefness of the exposure. If the flashed dots are made extremely bright they will leave an after-image on the retina. (This means that if you then move your eyes the pattern of dots appears to move with your eyes, being fixed on your retina.) You can see the pattern of dots for several seconds, but you still cannot count them accurately—a very strange feeling. You start to count, but lose track of which dots you have counted.

Is there a form of attention that does not depend on eye movements? Can attention move about between one large eye movement (called a "saccade") and the next? The American clinical psychologist Michael Posner at the University of Oregon has done many experiments on this.[1] He and others have shown that indeed there is such a form of visual attention. In a typical experiment the subject keeps his eyes fixed by gazing at a specific spot. A brief signal tells him that an object is likely to appear at a particular point, say to the right of his fixation point. He is asked to press a switch as quickly as possible when it appears. His reaction time is recorded. If, on some trials, it appears at an *un*expected place, such as to the left of the fixation point, his response is slower. This delay in response time is interpreted as being due to his having to shift his visual attention from the expected side to the unexpected side.

Posner has suggested that this change in attention is likely to involve three successive processes:

$$\text{disengage} \rightarrow \text{move} \rightarrow \text{engage}$$

First the system needs to disengage from the place in the visual field it was attending to. Then "attention" must be moved to the new place, and finally the system must engage attention there. Another important question is whether one can attend to two separate places (or objects) in the visual field at the same time. The balance of the evidence is that one cannot,* although one may be able to keep track of several† moving points.[3] There is substantial evidence, however, that attention can be finely focused in space or, alternatively, made more diffuse. For example, when reading a book you attend mainly to the words rather than to each separate letter. This will not do for proofreading, when you must scrutinize every letter and punctuation mark; otherwise small mistakes will easily be missed. Personally I find proofreading difficult. I normally read so fast that unless I make a special effort to focus my attention I don't notice tiny printing errors.

It is clear that attention alters the way you see. How do theorists attempt to explain this? Let me say straight away that, at the moment, there is no generally agreed theory of visual attention. The most I can do is to describe some of the current ideas and mention a few of the main points of controversy.

There is general agreement that, in some loose sense, attention involves a bottleneck. The basic idea is that early processing is largely parallel—a lot of different activities proceed simultaneously. Then there appear to be one or more stages where there is a bottleneck in information processing. Only one (or a few) "object(s)" can be dealt with at a time. This is done by temporarily filtering out the information coming from the unattended objects. The attentional system

---

*There is suggestive evidence that if the corpus callosum is cut, each half of the brain can attend to a different visual object.[2]

†It is possible, however, that the brain learns to regard these moving points as the corners of a single object that is changing shape.

then moves fairly rapidly to the next object, and so on, so that attention is largely serial (i.e., attending to one object after another), not highly parallel (as it would be if the system attended to many things at once).* I shall enlarge on this important distinction (between parallel and serial processing) later in the book.

A common metaphor is that there is a "spotlight" of visual attention. Inside the spotlight the information is processed in some special way. This makes us see the attended object or event more accurately and more quickly and also makes it easier to remember. Outside the "spotlight" the visual information is processed less, or differently, or not at all. The attentional system of the brain moves this hypothetical spotlight rapidly from one place in the visual field to another, just as, on a slower time scale, you move your eyes.

The spotlight metaphor in its simplest form rather implies that the visual system pays attention to one *place* in the visual field. There is much indirect evidence that it does this. An alternative is that attention is paid, not to a particular place, but to a particular object. If the object moves (the eyes being kept still) then in some cases attention can be shown to be attached to the object rather than staying in one place.[4] At the moment it seems likely that both forms of attention—to a visual object or to a visual place—can occur to some extent.

Psychologists have often made a distinction between preattentive processing and attentive processing. The Hungarian psychologist Bela Julesz, who has worked in the United States for many years, has given some dramatic examples of preattentive processing.[5] Look at Figure 20: The boundary between the two "textures" on the left leaps out at you. Now look at the right of the figure: At first glance one sees no obvious texture boundaries. Closer inspection shows that one region consists of the letter *L*, in various orientations; another region the letter *T*, but the difference does not pop out at you. Focal attention is needed to see it.

There is another way to study pop-out, or the lack of it. A visual display is flashed onto a screen and kept there. In this case the display often consists of a "target" that the subject has been asked to detect, located somewhere among other different but somewhat similar objects called "distractors." For example, a lot of letters may

---

*The brain can learn with practice to treat a particular set of "objects" (such as a set of letters) as a single "chunk," as it is called.

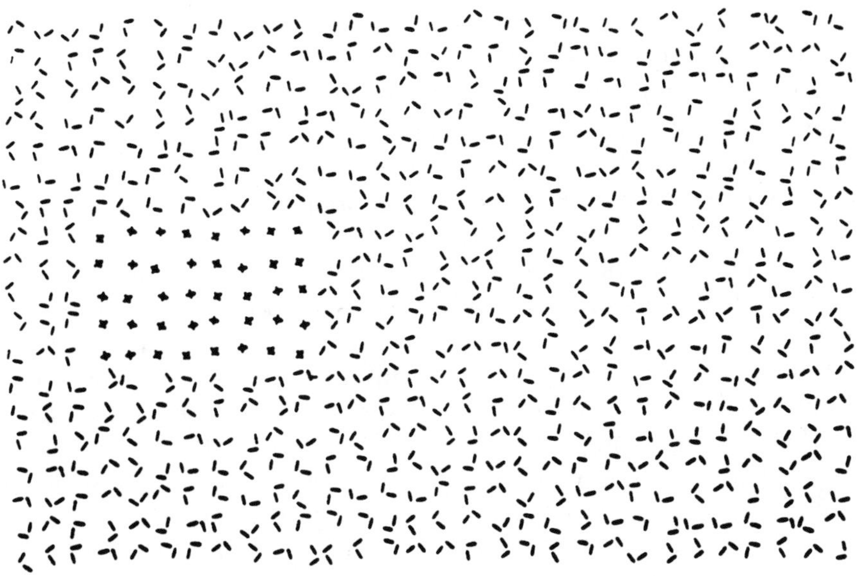

Fig. 20. How uniform is this?

be scattered over the display. All are green, except for a single red one. The subject is asked to press a button as soon as he spots the red letter. It is found that he can do this very rapidly. More important, his reaction time does not depend upon whether there are just a few green letters there, or many of them. It stays the same, no matter how many distractors there are. The red letter simply "pops out" at him.

Anne Treisman, one of the leading psychologists who study attention, did a famous experiment[6] with two colleagues in 1977. The gist of their experiments was this. It was first established that a red letter popped out against a background of green letters. It was also shown that when the letters were all the same color a single letter T popped out from a background of S's—that is, pop-out occurred for both shape and color. Then the subject was given a more difficult task. About half the letters were green T's, about half red S's, but in addition there was one red T. The subject was asked to look for this single red T. The subject could not just look for a single red letter, nor for a single T; there were too many of each of them. The subject had to look for the conjunction of a color (red) and a shape (T). This conjunction did not pop out at him. It took him some time to find the red T, and the more distractors there were, the longer it took. If there were twenty-five letters in the display, finding the single red *T* took much longer than if there were only five letters there.*

This was interpreted as evidence for a serial search mechanism—that is, the attentional system had to look *at one letter at a time* in order to decide whether that letter was both red *and* a *T*.

How much time does it take for attention to move from one place to another? Here things get more complicated. It seemed as if the more "salient" the object was—that is, made a big effect on the attentional system—the shorter the time. This may occur because if the red is very red (for example) the system may be able to examine *several* letters at once by making the "spotlight" spread over a larger area. This would mean that it needed to take fewer steps to look at all the letters, so that the processing time per letter was reduced. It was possible to argue that in dealing with a single object at a time it took about 60 milliseconds per step. For two objects at a time, the time per step was still 60 milliseconds, but the time *per letter* (which was all that one could observe) would now be only 30 milliseconds. For three objects at a time, only 20 milliseconds.

But there is a further complication. Perhaps the subject's brain learned to be clever and decided to pay attention only to red letters (and ignore the green ones).† Thus it could ignore about half the letters. This would mean it could complete the search more rapidly for the same step-rate of attention. In that case, a step-rate of 120 milliseconds would produce the observed result.

We thus have the unfortunate situation that in some cases the step-rate may appear to be as little as, say, 20 milliseconds even though its *true* rate may be as long as 120 milliseconds. This is because it may "cheat" by attending only to red objects and also by batching together three letters at a time until it finds the (red) *T*. Thus the correct time for each step of the spotlight is uncertain.

Treisman also showed that pop-out could be asymmetrical.[8] An interrupted circle will pop out against a background of complete circles (see Fig. 21a), but it needs serial search to find a complete circle against a background of interrupted ones (Fig. 21b).

How do psychologists describe the difference between preattentive processing and attentive processing? Originally, Treisman thought

---

*Response times vary quite a bit from trial to trial, so to get reproducible results it is necessary for the subject to make many responses and to take the average response time. In some cases several subjects are used, and their combined average response times are calculated.

†There is experimental evidence that this can happen.[7]

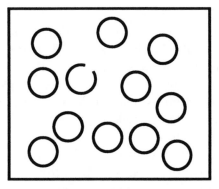 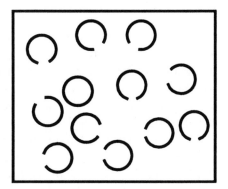

Fig. 21a. Spot the odd one.   Fig. 21b.

preattentive processing registered simple features (such as orientation, movement, color, etc.) in parallel, across the visual field, in a number of specialized subsystems. Focused attention then combined these features in some way. By careful experiment, she showed that if the time allowed for feature combination was made very brief, the brain might slip up and sometimes combine the features incorrectly, giving illusory conjunctions. As a demonstration, during her lecture, Treisman very briefly flashed a slide showing a young woman with black hair wearing a red pullover. Usually several people in the audience were convinced they had seen a woman with red hair. The red of her pullover had "migrated" to her hair, thus producing an illusory conjunction.

This can happen in everyday life, but very rarely. Treisman[9] gave an example: "A friend walking in a busy street 'saw' a colleague and was about to address him when he realized that the black beard belonged to one passerby and the bald head and spectacles to another."

What exactly is a "simple feature" is not known in advance.* Unfortunately much further work has shown that what pops out is not a straightforward matter—I shall not attempt to describe the details of the many experiments involved.

Treisman's various models of attention usually describe pop-out as being a distinct process from the longer process of serial search. Other psychologists, such as Kyle Cave and Jeremy Wolfe, have suggested

---

*This suggested a research program that explored which visual properties popped out—they would correspond to simple features (the "primitives" of vision)—whereas a compound of features would need serial search.

that pop-out is simply the first step in the process of attention.[10] They assume that the attentional system is somewhat "noisy" and therefore liable to make errors. If an object is sufficiently "salient," then the spotlight of attention will go to that place or object as its first step. If an object is much less salient, the system may have difficulty selecting the target object. It may make several attempts before it lights on the target, thus taking time. Such a mechanism can yield results that look like those of a simple serial search mechanism.

Even the existence of a spotlight has been denied by John Duncan and Glyn Humphreys.[11] They suggest that different objects in the visual field all start struggling to reach visual short-term memory and, if successful, can in some cases serve as a focus for action. Their model, which is hierarchical, also takes account of the relationship *between* different distractors—for example, whether the distractors are all the same or, alternatively, are of many different sorts.

Further work may allow psychologists to arrive at an agreed model of attention, although it is unlikely to be a simple one. I suspect that the correct model will not be derived from psychological testing alone—the system seems too complex for that. Some knowledge of the behavior of neurons in the relevant parts of the brain may be necessary before the correct answers emerge.

Visual attention, then, is only partly understood. At the moment we have no generally agreed psychological model for it.

What about short-term memory? What is known about that? A memory might be defined as a change in a system, due to experience, that makes some alteration to its subsequent thoughts or behavior, but this is too broad to be of much value. It would cover fatigue, injury, poisoning, and so on, and would not distinguish between learning and development (early growth). The Israeli neurobiologist Yadin Dudai has produced a more useful and more sophisticated definition.[12] He first describes what he means by an "internal representation" of the "world"—that is, of both the external and internal milieu. He defines an internal representation as "neuronally encoded, structured, versions of the world that could potentially guide behavior." This emphasizes that, at bottom, we are mainly concerned with how nerve cells (neurons) influence behavior. "Learning" is then the creation or modification of such an internal representation, produced by experience. Such changes persist for an appreciable time (sometimes for years), although we shall be interested mainly in memories that last only a very short time.

I shall not be concerned with very simple forms of memory, such as habituation or sensitization. (Suppose you show a baby the same picture ten times in succession. At first he is interested, but soon he becomes bored with it. This is called "habituation.") These processes are classed as "nonassociative." They occur even in very lowly animals, such as the sea slug. We shall be more interested in "associative learning" in which the organism responds to *relations* among stimuli and actions.*

It is useful to divide memory into several fairly distinct types, although exactly how they should be described is a matter of controversy. One convenient division is into episodic, categorical, and procedural memory. Episodic memory is a memory of an event, often together with irrelevant details associated with the event. A good example would be remembering where you were when you heard that President Kennedy had been assassinated.[†] An example of a categorical memory would be the meaning of a word, such as "assassination" or "dog," whereas the knowledge of how to swim or to drive a car would be classed as procedural memory.

Another method of classification depends on timing: how long it takes to acquire the memory and how long the memory usually lasts. Some memories, especially episodic memories, are classed as "one-shot" or "flash-bulb" learning. One remembers them strongly after only a single instance. (Such memories may of course be strengthened by rehearsal—by telling the story over again, not always correctly.) Other types of memory benefit from repeated instances, from which one extracts the general nature of something, such as the meaning of an (undefined) word.

Procedural knowledge, such as driving a car, is often difficult to acquire from a single experience and usually benefits from repeated practice. It often lasts for a remarkably long time. Once you have learned to swim you can swim fairly well even if you haven't swum for many years. A famous pianist said to me, about forgetting a familiar piece of music, "Muscle memory is the last to go," meaning by that term playing the piece automatically and without thinking about it.

Memories typically last for different times and are often divided

---

*There are other simple forms of memory that I shall not deal with here, among them classical conditioning, operant conditioning, and priming.

†Records show at an earlier period many people remembered vividly the occasion when they first heard that Abraham Lincoln had been shot.

into long-term and short-term memory, although the terms mean different things to different people. "Long term" usually means for hours, days, months, or even years. "Short term" can cover periods from a fraction of a second, to a few minutes or more. Short-term memory is usually labile and of limited capacity.

Consider what happens when you are dreaming. It appears that you cannot put anything into long-term memory (or at least, anything you can explicitly recall) while you are actually dreaming. Your brain holds the dream in some form of short-term memory. When you wake up (which may happen more often than you realize) the long-term memory system switches on. Anything still in short-term memory can then be transferred into long term—that is why you remember, not everything you have dreamt, but the last few minutes of the dream. If, shortly after waking, you are disturbed—by a telephone call, for example—your short-term memory of the dream, being interrupted, may decay and be lost, so that after the telephone call is over you can no longer recall even the last part of your dream.

Recalling a memory, as we all know, is not a straightforward process. Usually some clue is needed to address the memory, and even then the memory may be elusive. Some memories become weak, and need stronger clues to evoke them. Others appear to fade until they are completely lost. A related memory may intrude and block access to the one you want, and so on.

Consciousness in general, and visual awareness in particular, obviously incorporate into their processes much that we have already stored in long-term episodic and categorical memories. What concerns us more is very short term memory, since it is plausible to argue that if we lost *all* forms of memory for new events we would not be conscious. However, this essential form of memory need only last a fraction of a second or perhaps a few seconds at most. So let us concentrate on these very brief forms of memory.

Look at the scene in front of you, and then suddenly close your eyes. The vivid picture of your visual world vanishes very quickly, leaving you with only a dim recollection of it. This usually fades in a few seconds. Attempts to measure the time of fading have been made since the eighteenth century. A light moving in the dark, such as a glowing cigarette end, leaves a trail of light behind it. Modern studies of the length of such trails suggest that the perception of the light persists for about 100 milliseconds, although some of this may be due to after-images on the retina.

How do psychologists study the various forms of short-term memory? A classic experiment was done in 1960 by the American psychologist George Sperling.[13] He flashed a set of twelve letters on a screen for a very short time (50 milliseconds), arranged in three rows of four letters. The time was so short that the subject could remember only four or five of the letters. Then, in a second experiment, he asked the subject to report just one line of the display. He used a tone to indicate which line the subject was to read off, but this clue was not given until immediately *after* the visual display was switched off. In this case the subject could report about three of the four letters on the line indicated by the clue.

One might have expected, from this second experiment alone, that if the subject could report three out of the four letters on *one* of the three lines, he should have been able to report nine (three times three) of them on all three lines, whereas, as we saw, he could recall only about four or five out of the twelve. This strongly suggests that the letters are being read off by the brain from a visual trace that is rapidly decaying. Such a form of very short term visual memory is often called "iconic memory" from the word *icon,* meaning an image.

There have been many other studies of this. The time of decay is different if the visual field, before and after the display, is light or dark. With a dark field, the decay time is of the order of a second or so; with a lighter field, much less, perhaps a few tenths of a second. Such an effect of a bright background field is called "masking." One can also use patterns as a mask, but the two types of masking are very different. In brief, brightness masking occurs at an early stage in the visual system, possibly in the retina, before the information from the two eyes is combined. Patterned masking depends very much on the time interval between the presentation of the letters and the mask. The data suggest that it probably occurs at several levels in the visual system, after the information from the two eyes has been combined.

Iconic memory seems to depend on the persistence of a brief visual signal, not so much as from its time of *off*set but from the time of *on*set. This suggests that its biological function is to provide sufficient time (roughly in the 100–200-millisecond range) to allow for the processing of very brief signals, and this implies that *a certain minimum time is needed for adequate visual processing.*

There is also a slightly longer form of short-term visual memory. The English psychologist Alan Baddeley, who has studied this intensively,[14] calls this "working memory." A typical example is recalling a new seven-digit telephone number. The number of digits you can recall is called your "digit span." For most people it is usually about six or seven items. In other words, working memory has a limited capacity. Such memory seems to have several different forms, depending on the sensory input involved. For vision, he calls it "the visuo-spatial scratchpad." The times involved are typically several seconds. It also appears to be involved in visual imagery—when you try to recall a face or some familiar object. Its properties are sufficiently different from those of the shorter iconic memory that it is likely to involve different processes in the brain.

Is working memory necessary for consciousness? There is some evidence that suggests it is not. Some patients with brain damage have a very much reduced digit span—they can hardly recall anything except the last digit they heard—yet they seem otherwise normally conscious and, in fact, may have unimpaired long-term memory.[15] So far there have been no cases of a patient that has lost all forms of working memory, visual as well as auditory, but this may be because the damage that would produce such a defect, without causing other handicaps, has to be very precisely localized, yet in many different places, and so may never occur in practice.

Long-term memory seems different from either iconic or working memory. A subject shown about 2,500 fairly different color slides (for 10 seconds each) could still recognize about 90 percent of them after ten days. However, as the subject had only to recognize that he had seen a picture before (not recall it unaided, which is more difficult) he need have retained only a fraction of the information in each picture.

We shall not be much concerned with long-term episodic memory, since a patient with brain damage who cannot form new long-term episodic memories is still alert and conscious (see Chapter 12). It is short-term memory, and iconic memory in particular, that is likely to be intimately involved in the mechanisms of consciousness.

6

The Perceptual Moment: Theories of Vision

"Psychology is a very unsatisfactory science."
—Wolfgang Köhler

The times involved in the decay of iconic memory and of working memory can be fairly short. Can anything be said about the time needed for the various processes that lead up to or correspond to awareness? Recall that some cognitive scientists (Chapter 2), who like to describe the activities of the brain as carrying out computations, believe that what reaches consciousness is not the computations themselves but the *results* of those computations.

It has been claimed that certain brain activities do not reach consciousness unless they last for some minimum time.[1] If the activity is weak, this time may be as long as half a second. We need to know the sort of duration of activity that corresponds to a single "moment of perception," if only to guide our search for the neural correlate of awareness. What is the sort of time involved in a single processing period?

Consider the following case. A subject is shown a brief 20-millisecond flash of red light, followed immediately by a 20-millisecond flash of green light at exactly the same place. What does he report that he sees? He does *not* see a red flash followed rapidly by a green flash.

Instead, he says he saw a yellow* flash—just as he would have done if the two colors had flashed simultaneously. Yet if the green flash had *not* followed the red one, he would have reported a red flash. This means that he could not have been aware of the color (yellow) until after the information from the green flash had been processed.†

Thus, you do not experience directly the true onset of a stimulus. You can have no conscious estimate of the real duration of a short stimulus. As long ago as 1887, the French scientist A. Charpentier found that the longest flash that did not appear to be longer than a 7-millisecond control flash had a duration of 66 milliseconds.

Robert Efron, the American psychologist, wrote a very perceptive article[2] on this problem in 1967. He concluded that the duration of the processing period, estimated by several somewhat different methods, was about 60–70 milliseconds. This figure was for fairly simple stimuli that were salient enough to be easily observable. It would not be surprising if the processing period were longer than this for fainter or more complicated stimuli.

Can one say anything about the time needed for more complex processing? This usually involves presenting a visual stimulus, followed rapidly by a mask—that is, by a pattern, in the same place in the visual field, that interferes with some of the processes necessary to see the original stimulus. The detailed interpretation of such results is likely to be tricky. If the system is a simple, flow-through one, in which the signal progresses steadily from stage to stage, without pausing, and if the step to awareness takes no time, then the signals from the mask might never catch up with the signals from the stimulus. Since masking usually interferes with the perception of the stimulus, this implies that at least some of the processing steps take time, which is in any case plausible. In spite of these difficulties in interpretation, the effects of masking can give us some limited information about what is going on.

The American psychologist Robert Reynolds has done several experiments[3] to investigate this. He wanted to show that different aspects of a percept could be seen at different times. In other words he want-

---

*The mixture of a red pigment and a green pigment gives a brown pigment, but the mixture of red light and green light produces yellow light.

†There are philosophical quibbles about whether he could have been briefly "aware" of the red flash but rapidly forgot it completely. This is clearly not the usual usage of "awareness." Such questions are best postponed until we understand more exactly what goes on in the brain under such conditions.

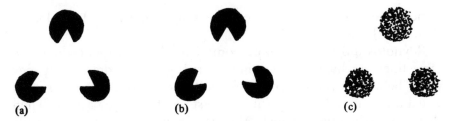

Fig. 22. A straight white triangle, a curved white triangle, and a mask.

ed to study the time course of the percept from the time the stimulus was presented until a relatively stable percept developed.

As an example, consider the time taken to develop the percept of illusory contours described in Chapter 4. Reynolds showed subjects one of two different examples of this (see Fig. 22)—making it more difficult for the subject to guess or lie. Each had three black "pacmen," so drawn that the sides of the illusory triangle appeared straight in the first one and curved in the second. A stimulus was flashed for 50 milliseconds, followed after a delay* by the mask shown in Figure 22c. The stimulus was sufficiently large and bright that even with this brief presentation the subjects always saw the three pacmen clearly. Because of iconic memory, we would expect that, without masking, the signals from the display might affect the brain for more than the 50 milliseconds for which it was flashed—probably for several hundred milliseconds.

Reynolds found that if the mask followed immediately after the stimulus, the great majority of the subjects did not see an illusory triangle, and the few that reported that they did often made errors, mistaking a straight triangle for a curved one, or vice versa. However, if the delay was 50–75 milliseconds—that is, at an SOA of 100–125 milliseconds—then all the observers reported seeing a triangle, though they were not all completely accurate about its curvature.

This clearly shows that the total processing time depends on exactly what one sees. The three pacmen could be clearly seen some time before the illusory triangle was seen.

Notice that these experiments do not show exactly at which time the "neural correlate" of the percept arose in the brain. Only that for

---

*Reynolds reported his results in terms of stimulus onset asynchrony (SOA). Since the stimulus lasted for 50 milliseconds, an SOA of 50 milliseconds means that the mask started immediately after the stimulus finished. I have called this zero delay.

some aspects of the percept the processing time is likely to be longer than for others.

Reynolds did another, more complicated, experiment along the same lines in which the same illusory triangles were drawn as if they were "behind" a transparent brick wall. The interpretation of such a visual image is ambiguous. Subjects first saw the three pacmen. Then they reported a bright triangle, then this triangle was rejected. Finally the percept of the triangle returned.* The time for the three latter stages was about 150 milliseconds each.

Clearly, the timing of "computations" can depend on their complexity. Even though a detailed interpretation will depend upon exactly how the signals move around different areas of the brain, and how these areas interact together (which is unlikely to be simple), at least we have a rough idea of the sort of times needed for visual processing. We are unlikely to get more accurate times until we understand better the many different brain processes involved in seeing and the way they interact.[†]

I have dealt briefly with quite a number of aspects of visual processing, but I have given no systematic account of how we should think about all these processes. This is a difficult subject. If this were a book solely about visual perception I should have to describe, at some length, some of the current ideas about vision—that is, how the brain actually carries out the complex activities that lead up to what we see. Apart from the cognitive scientists mentioned in Chapter 2, most of these theorists have shown little interest in consciousness. Because of this, and because there is no generally agreed theory of vision, I have omitted a detailed description of the many rather different approaches to the problem. The following brief overview, however, should give the reader a fleeting glimpse of the subject.[‡]

People are interested in vision for very different reasons. Some want to build a vision machine that can see as well as we do, or better, so that it can be used for domestic, industrial, or military purposes. They are not much concerned with how our brains do the job, except as a source of ideas. A vision machine need not imitate the brain exactly, any more than an airplane has to flap its wings.

---

*Note that the subjects did not report all these stages in a single experiment. The results are deduced by comparing the reported percepts after different delays of masking.

†I have postponed consideration of some of Libet's work to Chapter 15.

‡Of course, for those doing experiments on visual awareness, it is essential to have a fairly detailed knowledge of the psychology of vision and of the various theories of visual perception, if only to prevent avoidable errors.

Others are mainly interested in how human beings see things. At one extreme are some of the Functionalists who deny that knowing about the details of the brain will ever tell them *anything* useful.* This point of view is so bizarre that most scientists are astonished to learn that it exists. At the other extreme are those neuroscientists who concentrate mainly on how nerve cells in an animal's brain respond to a visual image and have little concern with how this activity results in seeing. Fortunately, there is now a handful of students of vision who fall in between these two extremes. They are interested in both the psychology of vision and also in how nerve cells behave.

The ideas people bring to these problems are also very different. Some think the important thing to study is the visual environment—the ground beneath us, the sky above us, and the manifold objects in between. They do not worry about the brain since they think that all it need do is "resonate" to aspects of the environment, whatever that means. They call themselves Gibsonians, after their guru, the late J. J. Gibson. Others try to analyze basic but rather limited visual operations, such as shape-from-shading, or the barber's pole illusion, and invent computer routines that will solve such problems. This tradition is still very strong in departments of Artificial Intelligence (A.I.). Still others liken what goes on in the brain to objects or events in everyday life. They talk about "spotlights" or "starting a file for an object." In the last twenty or thirty years the type of explanation used has often been based on how a computer does things, using a set of explicit rules to arrive at a conclusion and involving such computer concepts as central processing, random-access memory, and so on. A more recent development is that of neural networks—sets of interacting neuron-like units that interact largely in parallel, without explicit rules. (This is discussed more fully in Chapter 13.)

The Gestalt psychologists, as we saw in Chapter 4, wanted to discover the basic principles underlying the act of seeing. It is argued that since it is necessary to know the laws of aerodynamics to understand both birds and airplanes, it follows that to understand vision one should look for the general principles involved. Modern versions of this approach often couch their theories in informational terms. Mathematicians, not surprisingly, tend to invent general mathematical principles of one sort or another. It would take a whole book to describe all these ideas to the general reader.

---

*"All you need to know about brains is how to cook them." Philosophers, A.I. workers, and linguists often adopt this point of view, but it is not unknown among refugees from the more exact sciences.

All these points of view have some value, but they have not yet been welded together to produce a detailed and widely accepted theory of vision. All present theories are inadequate, if only because they do not face up to the problem of visual awareness. Vision is in any case so complex and so difficult a process that we are unlikely to get a comprehensive theory of it till some time in the next century. If we wish to tackle the problem of visual awareness now we shall have to get along as best we can. To do this we need some tentative point of view, or we shall get lost.

One approach I have found useful was put forward by the late David Marr. David was a young Englishman who took a degree in mathematics at Cambridge University as a preliminary to studying the brain. His Ph.D. thesis presented a detailed and novel theory of the cerebellum. Subsequently, Sydney Brenner and I gave him an office in our laboratory at Cambridge, England, where he produced theories about the general operation of the hippocampus and the visual cortex. He became a partial convert to the A.I. approach to vision, and moved to the Massachusetts Institute of Technology (MIT), where he collaborated with the Italian theorist Tomaso Poggio. They both visited me at the Salk Institute for a month in April 1979. David wrote a book entitled *Vision* (published posthumously) that explained in a straightforward way—his scientific papers are not easy to read—his many seminal ideas about vision. Not all these have stood the test of time, but the book is still a masterly exposition of the problems as he then saw them. The last chapter consists of an imaginary dialogue between David and a reluctant believer (myself), largely modelled on the many conversations the three of us had while he and Tommy Poggio were at the Salk.

David had conceived a general scheme that described the broad outlines of the process of seeing something. He believed that the main job of vision was to derive a representation of shape; brightness, color, texture and so on he considered secondary to shape. He naturally took the view that the brain builds within itself a symbolic representation of the visual world, making explicit the many aspects of it that are only implicit in the retinal image. David thought, almost certainly correctly, that the job could not be done all in one step. Instead he postulated that there has to be a *sequence* of representations. He called these the "primal sketch," the "2½D sketch," and the "3D model" representation.

The *primal sketch* makes explicit some important information about the two-dimensional image, primarily the intensity changes there and their geometrical distribution and organization. It deals, among other things, with edge segments, blobs, terminators, discontinuities, boundaries, and so on. The *2½D sketch* makes explicit the orientation (and rough depth) of the visible surfaces and their contours, in a viewer-centered frame. The *3D model* representation describes shapes and their spatial organization in an object-centered frame.

This at least divides the visual task into separate stages and is of some use if only because it makes us realize how much has to be done in seeing something. It is unlikely to be correct in detail. The three stages are probably only a first approximation—for example, color, texture, and motion must be added to "shape." There may be more than three stages and they are unlikely to be as distinct as his description implies. They probably interact in both directions. Nevertheless, his scheme does suggest the sort of processing that may be taking place when you see something. (I discuss its relevance to neuroscience in Chapter 17.)

David Marr's early death, at thirty-five, from leukemia was a great loss to theoretical neurobiology. I am convinced that, had he lived, he would not have remained fixed in his ideas but would have developed further approaches to brain theory as the subject progressed. His incisive mind and his imaginative creativity would surely have helped us through the tangle of difficulties that confront us today. He combined very considerable intellectual powers with the ability to absorb and digest a large amount of experimental evidence of many different kinds.

What style of explanation shall we need to understand the brain? My own view agrees most closely with V. S. Ramachandran's Utilitarian Theory of Perception. He argues that visual perception does not involve intelligent deduction of exactly the type we use in constructing an argument, nor does it involve the vague idea that the brain simply "resonates" to the visual input. Neither does it require the solving of elaborate equations, as often implied by A.I. researchers. Instead he believes perception "uses rules of thumb, short-cuts, and clever sleight-of-hand tricks that are acquired by trial and error through millions of years of natural selection. This is a familiar strategy in biology but for some reason it seems to have escaped the notice of psychologists who seem to forget that the brain is a biological organ. . . ." I also agree with Ramachandran when he states: "The best way to resolve some of these issues may be actually to open the black box in order to study the responses of nerve cells, but psychologists and computer scientists are often very suspicious of this approach."[4]

In Ramachandran's view, the job of the visual psychologist is not, at this stage, to construct elaborate mathematical theories to explain his results but instead to sketch what might be called the "natural history" of vision, especially the earlier stages of vision. When the visual task has been dissected into its many component parts, and especially when it can be shown that certain interactions are weak or absent, we shall know just what needs to be explained in neural terms. These explanations may or may not involve elaborate mathematics. They will certainly involve the properties of interacting neurons and the details of their interconnections. Thus, because of the complexity of the visual world, one expects to find many rapid, rough-and-ready processes interacting dynamically in many different ways.*

The next step, then, is to learn something about the human brain (and the monkey brain) and about the many nerve cells and molecules of which they are made. This is the subject of Part II.

---

*There may be only a few basic *learning* mechanisms underlying all this complex activity. The final explanation is likely to be in terms of the basic patterns of connections laid down in normal development, plus the *key learning algorithms* that modify those connections and other neural parameters. Thus, the neocortex may well have an underlying simplicity, not at the level at which the mature brain behaves but at the way by which it arrives at that intricate behavior, based on its innate structure and guided by its rich experience of the world.

Part II

Part II

7

The Human Brain
in Outline

"And still they gaz'd and still the wonder grew,
That one small head could carry all he knew."
—Oliver Goldsmith, *The Deserted Village*

The nervous systems of all mammals, from mice to men, are
built according to the same general plan, although they can differ greatly in size—compare a mouse's brain to that of an ele-
phant—and also in the proportions of the various parts. The brains of
reptiles, birds, amphibians, and fish are clearly related to the brains of
mammals but there are significant differences. I shall not consider
them further here. Nor shall I attempt to describe the brain's development in the fetus and in the young animal, although this is an important topic that can help us to understand the mature brain. It suffices
to say that genes (and the epigenetic processes they control during
development) appear to lay down the broad structure of the nervous
system, but that experience is needed to tune up and refine the many
details of its structure; this is often a continuing process throughout
life.

There is one fact about the brain that is so obvious it is seldom
mentioned: It is attached to the rest of the body and communicates
with it. The nervous system receives information only from the various

transducers in the body. (A "transducer" turns a chemical or physical influence, such as light, sound, or pressure, into an electrochemical signal.)

Some of these transducers respond to signals that come largely from outside the body, as the photoreceptors of the eyes do when they respond to light. They monitor the external milieu. Other transducers respond to activity that is largely inside the body, such as those that react when you have a stomach-ache or are sensitive to the acidity of your blood. They monitor the internal milieu. The motor output of the nervous system controls the many muscles of the body. The brain can also influence the internal release of various chemicals, such as certain hormones. The peripheral nerve cells that are directly concerned with all these inputs and outputs are only a very small fraction of the total. The great majority of nerve cells process information *inside* the system.

The central nervous system can be divided in different ways but a simple division is into three parts: the spinal cord, the brain stem (at the top of the spinal cord), and the forebrain above that. The spinal cord receives sensory information from the body and transmits instructions to its muscles. Since we are concerned with vision, neither the spinal cord nor the lower part of the brain stem require much further attention. Our main interest will be with the forebrain and, in particular, the neocortex, the largest part of the cerebral cortex.

The cerebral cortex (usually called simply the cortex) consists of two separate sheets of nerve cells, one on each side of the head. For humans, the total area of these two sheets is a bit more than that of a man's handkerchief. For this reason it has to be extensively folded to fit into the skull. The sheet varies somewhat in thickness but is typically 2–5 millimeters thick. This constitutes the gray matter of the cortex. It consists mainly of neurons,* their bodies, and their branches, although there are also many accessory cells, called "glial cells." There are about 100,000 neurons beneath every square millimeter of the cortical sheet,[†] so that altogether the human neocortex contains some tens of billions of neurons, comparable in number to all the stars in our galaxy.

Some of these connections between neurons are local—they only go

---

*As mentioned in Chapter 1, "neuron" is the scientific word for a nerve cell.

[†]Except for the first visual area of primates, which have rather more than twice as many.

a fraction of a millimeter or, at best, a few millimeters—but others leave the cortical sheet and travel some distance before entering another part of the sheet or going elsewhere. These longer connections are often covered by a fatty sheath, made of a material called myelin, that enables the signal to travel faster and gives this tissue a somewhat white, glistening appearance, so that it is called "white matter." About 40 percent of our brain is made of white matter—that is, of these longer connections. This demonstrates, in a simple and graphic manner, how much communication there is within the brain.

The neocortex is the most complex part of the cortex (see Fig. 23). The "old" cortex (the paleocortex), also a thin sheet, is mainly concerned with smell. The hippocampus (sometimes called the "allocortex") is an interesting high-level structure (meaning that it is far from the sensory inputs). It probably stores, for a few weeks or so, the codes for new, long-term, episodic memories before the information is conveyed to the neocortex.

Associated with the cortex are various subcortical structures inside the fore part of the brain (Fig. 23). The most important of these is the

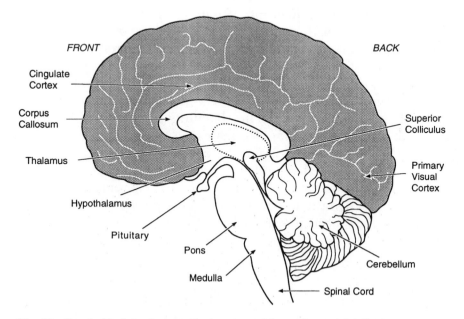

Fig. 23. One half of the human brain, viewed from the inside, showing various major parts. The cortex is shaded gray.

front   back

THALAMUS

Fig. 24. The key position of the thalamus and its connections to (and from) the cerebral cortex.

thalamus,* sometimes called the gateway to the cortex, because *the main inputs† to the cortex have to pass through it* (Fig. 24). The thalamus is conveniently divided into about two dozen regions, each of which is concerned with some particular subdivision of the neocortex. Each thalamic area also receives massive connections from the cortical areas to which it sends information. The exact purpose of these back connections is not yet known. The thalamus does not stand in the way of many other connections *from* the neocortex. These can go to other parts of the brain directly. The thalamus sits astride the main entrances to the cortex but not the main exits.

Near the thalamus (Fig. 25) are certain well-developed structures, usually grouped together under the term *corpus striatum*. These regions play an important role in the control of movement although their exact function is unclear. Special regions of the thalamus (called collectively the "intralaminar nuclei") project mainly to the corpus striatum and also, more diffusely, to the neocortex.

*The word *thalamus* comes from a Greek word meaning an inner room commonly applied to a bridal chamber or bridal couch. A large part of the visual thalamus is called the "pulvinar"—the word originally meant a pillow!

†This is not true for certain somewhat diffuse systems from the brain stem and elsewhere.

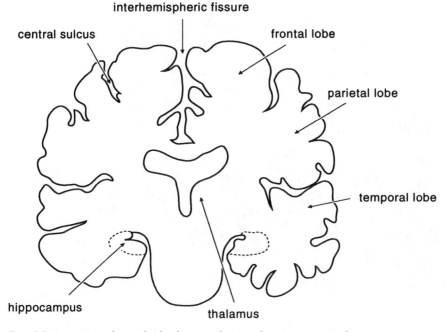

Fig. 25. A section through the human brain, showing certain key parts.

*   *   *

There has been continual controversy over the last hundred years or so about where different mental functions are localized in the neocortex. At one extreme is the holistic view that all parts of the cortex are broadly equivalent in function, while at the other extreme is the opinion that each small region of the cortex performs a quite different task.

In the early part of the last century, the Viennese anatomist Franz Joseph Gall believed in localization and labelled the parts of the skull with various fanciful attributes (e.g., sublimity, benevolence, and veneration) that were supposed to be located in the underlying cortex (Fig. 26). Ceramic models of the human head, with these labels written on them, still exist. Gall believed that by studying the bumps on the skull he could deduce a lot about a person's character. When I was a boy a local charlatan persuaded my mother to give him money to "read my bumps." He claimed that my bumps were of great interest, and additional money was extracted so that he could study them more closely. What he deduced about me from all this I never discovered.

Although Gall was the first important advocate of localization, his detailed ideas about it were quite wrong, and as a result, cortical local-

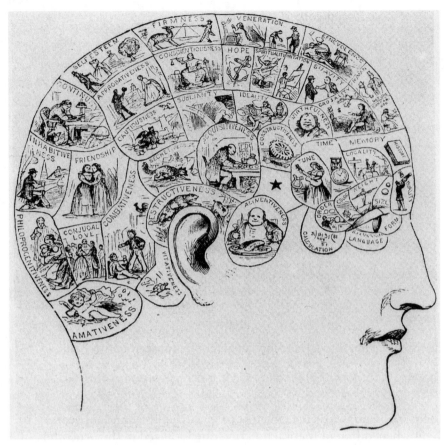

Fig. 26. A wholly imaginary nineteenth-century view of the localization of function in the human brain, based on the ideas of Gall.

ization got a bad name in medical circles. Now, largely through detailed studies on the cortex of the macaque monkey, supported by data from humans, we know that *some* degree of localization does occur. However, many distinct cortical areas have to participate cooperatively during most mental activities, so the idea of localization must not be taken to extremes.

Perhaps a useful analogy would be the properties of a small organic molecule, such as sugar or vitamin C. The position of each atom is localized with respect to the others. Each distinct atom has characteristic properties of its own—for example, oxygen is very different from hydrogen. The overall behavior of the molecule, however, depends on how the constituent atoms interact with each other, even though

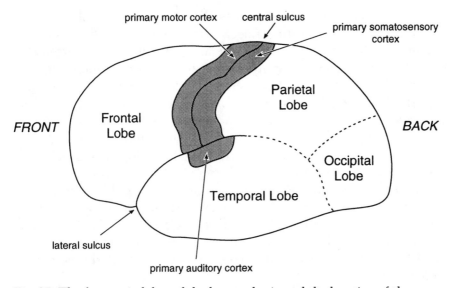

Fig. 27. The four main lobes of the human brain and the location of the main motor area and primary sensory areas.

some of the atoms are usually more important than others. Sometimes the electrons that link the atoms together are fairly localized. In other cases, as in aromatic compounds like benzene, some of them are distributed over a number of atoms.

We can therefore make a rough map of the neocortex, and label the different regions according to their primary function (see Fig. 27). Vision is located at the back of the head (see Fig. 23), hearing at the side, and touch at the top. Just in front of the somatosensory region (touch) lie the areas that control the voluntary motor output— that is, the willed instructions to the muscles. The exact functions of the frontal regions are less certain. They probably deal with planning, especially long-term planning, and other high-level cognitive tasks. A small area in the frontal region (the frontal eye fields) seems to be involved in voluntary eye movements.

It is commonly known that, oddly enough, the *left* side of the cortex relates most directly to the *right* side of the body.* However, a massive tract of nerve fibers, called the "corpus callosum," joins the two cortical

---

*Smell is an exception. The right side of the nose connects to the right side of the brain.

sheets together. In man, the corpus callosum has about half a billion individual nerve fibers, with connections running in both directions.

Language, which is unique to humans, is mainly dependent on the left side of the head for virtually all right-handed people and most left-handers. There are at least two main regions. One of them, at the side near the back, is called "Wernicke's area." The other, the first to be discovered, called "Broca's area," is at the side, nearer the front, not far from the main motor region. Neither region is understood in detail, largely because there is no animal with highly developed language, and animals are our main source of experimental information about the brain. There are several other regions in the vicinity of these two, especially in the temporal part of the cortex, that are also involved in processing language (see Chapter 9). I believe that each of these large regions, including Broca's area and Wernicke's area, will turn out to consist of many distinct smaller cortical areas, connected together in complicated ways.

A stroke on the left side of the head, if it is extensive enough, may paralyze parts of the right side of the body and also interfere with speech. Yet the undamaged right-side brain may still be able to swear and even to sing. In addition, such a person may still be able to distinguish a male voice from a female voice. This latter ability may be lost if there is damage to the *right* side of the brain, which leaves speech largely intact, although the music of speech may be lost.

These examples illustrate two points: There is indeed some degree of localization in the brain, but *what* is localized may not be what one would have guessed.

A region known as the hypothalamus (see Fig. 23), outside the cortex, is essential to many operations of the body. It has many small subregions whose main functions are to regulate hunger, thirst, temperature, sexual behavior, and similar body operations. The hypothalamus has a close connection with the pituitary, a tiny organ that secretes various hormones into the bloodstream.

A much larger, more striking but less essential brain region is the cerebellum (Fig. 23), at the back of the head. This is highly developed in certain fish, such as electric fish and certain sharks. It appears to be involved in the control of movement, and in particular with the efficiency of skilled movement. Nevertheless, a person born without a cerebellum can live reasonably well without it. Another important region, broadly located in the brain stem, is called the "reticular formation." This has various, closely interacting parts, the functions of which are only partly understood. It is here that the nerve cells that control general arousal and the various stages of sleep are found. Groups of these nerve cells send signals to various parts of the forebrain, including the neocortex. For example, a small group of neurons, called the "locus ceruleus," sends signals to various places, including the cortex. A single one of these nerve fibers can extend from the front of the cortex to the back, making millions of connections to other nerve cells on the way. The exact function of the locus ceruleus is unknown. It becomes almost completely inactive in the rapid eye movement (or REM) phases of sleep during which most of our dreams occur. It is possible that its activity is needed when the cortex needs to put a memory into long-term storage. Its inactivity in REM sleep may help to explain why we are unable to recall the majority of our dreams.

A pair of structures at the top of the brain stem are important to the visual system. These are called the "optic tecta" in lower vertebrates, such as frogs, and the "superior colliculi" in mammals. They probably constitute a major part of a frog's visual system, but in mammals (and primates in particular) they have yielded this role to the neocortex. In mammals, the main concern of the superior colliculus appears to be with eye movements, and especially involuntary eye movements.

The human brain is not a uniform structure, any more than the rest of our body is. Just as heart, liver, kidneys, and pancreas have distinct functions, so do the various regions of the brain. Yet the distinct organs of the body can interact quite closely together. The liver is "the organ of the blood" but the heart pumps the blood. Many interactions are also found in the brain. The control of movement often involves not only the spinal cord but many regions above it, such as the motor cortex, the corpus striatum, and the cerebellum. Vision involves both the superior colliculus and the visual parts of the thalamus and visual cortex, though the jobs they have to do are somewhat different.

We understand fairly well the main functions of almost all of the body's organs and also, in a broad way, how each organ carries out these functions. In one or two cases this knowledge is fairly recent. When I started biological research in the late forties the function of the thymus was not known and it was not even suspected that it played a key role in our immune system. Indeed, I first learned of it because calf thymus was a convenient source of DNA. Unfortunately, our knowledge of the different parts of the brain is still in a very primitive stage. What *exactly* are the functions of the thalamus, or the corpus striatum, or the cerebellum? We can glimpse the general outlines of their behavior, but detailed knowledge has yet to come. We have a very rough idea of what the hippocampus does, but there is no agreement as to its exact function. All this remains to be discovered.

Having seen what the brain is like at the highest level of description let us now dive down to a much lower level and take a look at its key components, the individual nerve cells.

8

The Neuron

"The function of the brain cannot be completely disconnected from the function of its basic units, the nerve cells."
—Idan Segev

Since the Astonishing Hypothesis stresses that "You" are largely the behavior of a vast population of neurons, it is important for you to have at least a rough idea of what neurons are like and what they do. Despite the fact that there are many distinct types of neurons most of them are built according to a common blueprint.*

A typical vertebrate neuron responds to the many sources of electric impulses that impinge on its cell body and on its branches—its "dendrites" (see Fig. 28)—in three ways. Some inputs excite the neuron, some inhibit it, and others modulate its behavior. If the neuron becomes sufficiently excited, it responds ("fires") by sending an electrical pulse—(a spike)—down its output cable—(its axon). This single axon usually branches. The electrical signal travels down each branch and subbranch, until eventually the axon contacts many other neurons and so influences their behavior.

This, then, is the major job of a neuron. It receives information, usually in the form of electrical pulses, from many other neurons. It does what

---

*I will concentrate on the "typical" neuron found in vertebrates like ourselves, as those in invertebrates (such as insects) are a little different.

91

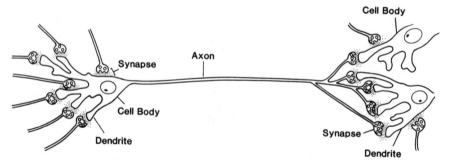

Fig. 28. A schematic diagram of a "typical" vertebrate neuron. The electrical signals flow into the dendrites and out through the axon. Thus, in this diagram, the information flows from left to right.

is, in effect, a complex dynamic sum of these inputs, and as a result sends out information in the form of a stream of electrical impulses down its axon and on to many other neurons. Although a neuron requires energy to sustain these activities and to synthesize molecules, its main function is to receive signals and to send them out—that is, to handle information. In a similar way, a politician is constantly bombarded by information from people wanting him to vote for or against a measure. He has to take all this information into account when he records his vote.

When nothing much is happening a neuron usually sends spikes down its axon at a relatively slow, irregular, "background" rate, often between 1 and 5 Hertz (1 Hertz is one spike, or one cycle, per second). This continual "nervy" activity keeps the neuron alert and ready to fire more strongly at a moment's notice. When it becomes excited, because it receives many excitatory signals, its rate of firing increases to a much higher rate, typically 50–100 Hertz or more. For short intervals the firing rate may reach 500 Hertz (see Fig. 29). Five hundred spikes a second may sound fast, but it is abysmally slow compared to the processing speed in a home computer, which can easily be a million times faster. If a neuron receives an excess of inhibitory signals, its output of spikes may be even less than its normal background rate, but this diminution is so small that it can convey rather little information. Neurons can send signals of only one type down their axons. There are no "negative" spikes. Moreover, these electrical signals normally flow only one way down the axon, from cell body to axon terminals.*

*   *   *

*By artificial means, the signals can be made to travel in the opposite direction, called "antidromic."

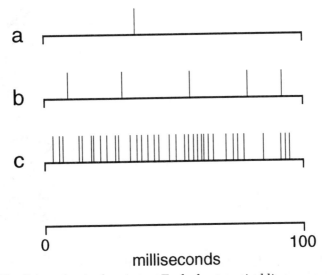

Fig. 29. The firing of a single neuron. Each short vertical line represents a spike. In *a*, the neuron is firing at its background rate. In *b*, it is responding to some relevant input by firing at an average rate. In *c*, it is responding about as fast as it can. Notice the time scale.

But what is a neuron like in detail? Of what is it made? In many ways it is just like other cells in the human or animal body. Its many genes are made of DNA, which comes packaged in the form of chromosomes that are contained in a special structure inside the cell called the "nucleus." There are other special structures, some of which (such as the mito-chondria, "the powerhouse of the cell") have their own DNA. Almost every cell in the body* contains two copies of the genetic information, one derived from each parent. Each set has a large number of distinct genes, probably about 100,000 of them.[†] Not all of these genes are active in all our cells. Some are more active in the liver, some in the muscles, and so on. It is thought that more genes are active in the vari-ous parts of the brain, taken together, than in any other organ.

Most of these genes code the instructions for the synthesis of one protein or another. If we think of each cell as a factory, then proteins are the delicate and rapid machine tools that make the factory work. The volume of a "typical protein" is usually appreciably less than a bil-

---

*Red blood cells are one exception.
[†]A more exact number is not yet known but is likely to be by the year 2000.

Fig. 30. One important type of neuron—a pyramidal cell. This drawing was made by the Spanish neuroanatomist Cajal about a hundred years ago.

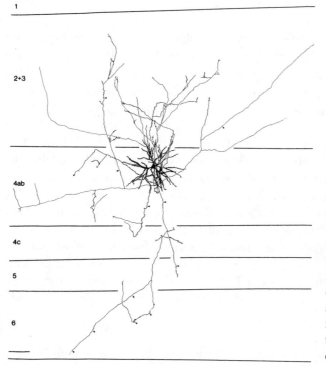

Fig. 31. Another type of neuron—called a "spiny stellate." The thin lines show the many branches of its axon. The thick lines represent its dendrites. The numbers on the left identify the different cortical layers, as if we were viewing the cross-section of a slice of the cortex.

lionth of the cell's volume and is thus far too small to be seen in the optical microscope, although its shape (but not the fine details of its approximate atomic structure) can sometimes be seen using an electron microscope. Each type of protein has its own precisely defined molecular structure, consisting of thousands, or tens, or even hundreds of thousands of atoms, all connected in a characteristic manner. The key molecules of life are constructed with atomic precision.

The entire content of the cell is enclosed in a somewhat fluid lipid (or partly fatty) membrane that prevents all this elaborate machinery and its products from leaving the cell. Certain proteins in this membrane act as delicate gates or pumps that allow various molecules to enter or exit the cell. The whole structure has elaborate control mechanisms, also made of organic molecules, that permit the cell to reproduce itself and to interact efficiently with other cells in the body. In short, it is a chemical miracle in a very small space, evolved by natural selection over billions of years.

Neurons are rather different from many of our other cells. Mature neurons do not move about, nor do they normally round up and divide. If a mature neuron dies, it is (with rare exceptions) not replaced by a new one. Neurons have a more spikey shape than most cells. The general style of dendritic branching varies from one type of neuron to another but a neuron usually has several main branches, each of which branches a few times into subbranches and so on. The body of the cell (often called its "soma") comes in various sizes. A typical diameter might be 20 microns.*

The most common type of neuron in our neocortex is called a "pyramidal cell." It often has a somewhat pyramid-shaped cell body, leading to a large apical dendrite—see Figure 30. Other neurons, such as "stellate" neurons, have branches in all directions (see Fig. 31).

The axon of a neuron (its output cable) can be very long—as much as several feet long in some instances, such as your spine, otherwise you'd not be able to waggle your toes. (Remember that the radius of the cell body of a neuron is seldom as large as a thousandth of an inch.) The diameter of an (unmyelinated) axon is usually very small—in the range of 0.1–1 microns, although some axons have a fatty (myelin) sheath that allows the spike of electrical activity to travel faster than it would without the sheath.

---

*Its volume is about a thousand times larger than a cell of a bacterium, such as *E. coli.*

*   *   *

The spike in an axon is not like an electric current in a wire. In a metal wire, the current is carried by a cloud of electrons. In a neuron, the electrical effects depend upon charged atoms (ions) that move *in or out* of the axon through molecular gates made of protein, in the insulating membrane of the cell. As the ions move back and forth they make local alterations to the electrical potential (or voltage) across the cell membrane. It is this change of potential that is propagated down the axon. This signal is a regenerative one, requiring energy. As a result, the spike travelling down the axon does not decay but stays approximately the same shape and size at the end as it was at the beginning. This property allows the spike to travel long distances and still produce an appreciable effect on the neurons it contacts at the end of its axon.

In the nineteenth century, it was assumed, quite erroneously, that the spike travelled at speeds too fast to be measured, possibly at the speed of light. When the speed was finally measured by Helmholtz in the middle of the last century, it was found to seldom be as fast as 300 feet per second. (This is about a third of the speed of sound in air.) Many people, including Helmholtz's father, were very surprised by this result. A more typical speed, for an unmyelinated axon, might be 5 feet per second. This may seem rather slow (slower, in fact, than a bicycle), but it is equivalent to 1.5 millimeters in about a millisecond.

The far ends of the axon have to be supplied with molecules by the cell body because almost all the genes and most of the biochemical machinery needed for the synthesis of proteins are in the cell body, not in the axon. There is a systematic flow of molecules along the axon in both directions. It is quite remarkable to watch a (speeded-up) movie of this, taken with the aid of a special high-powered optical microscope, showing minute particles jogging past each other, some going down a large axon, some going up it. Some travel faster than others, but all these flows are far slower than the speed of the axonal spike. Naturally, special molecular apparatus is needed to direct and to power this transport.

The classical view of a neuron held that the dendrites (the input cables) were "passive." This implies that the change of potential decayed as it spread from one position on the dendrite to another, as some of the ions involved leaked through the membrane, just as Morse signals used to decay as they travelled great distances along transatlantic cables. For this reason, dendrites are typically shorter than axons, often being only a few hundred microns long. It is now suspected that some neurons have active processes in their dendrites (as axons have), but they are probably not exactly the same as those found in axons.

The spike, then, travels down the axon till it reaches a synapse, the special junction between one neuron and another. Each neuron has many synapses on its dendrites and soma. A small neuron may have as few as five hundred; a large pyramidal cell may have as many as twenty thousand. An average number for neurons in the neocortex might be six thousand. It might be thought that since the spike is electrical, and the effect on the next neuron is mainly electrical, that the synapse is some form of electrical contact. This is sometimes the case, but more usually the transmission from one neuron to the next is more complicated than that.

In fact, the two neurons are not directly joined together. There is a well-defined gap between them of about one-fortieth of a micron, easily visible in pictures taken with an electron microscope (see Fig. 32). This gap is called the "synaptic cleft." When the spike arrives at the (presynaptic side of) the synapse, it causes little packets of chemicals (called "vesicles") to be released into the gap. These small chemical molecules diffuse rapidly in the gap, many of them combining with one or more of the molecular gates in the membrane of the synapse of the recipient cell. This causes those particular gates to open and allows charged ions to flow in or out of the membrane on the postsynaptic side of the synapse, so that the local potential across that membrane is changed. The overall process is:

$$\text{electrical} \rightarrow \text{chemical} \rightarrow \text{electrical}$$

Whether ions flow in or out depends, loosely speaking, on whether their concentration is higher or lower inside the neuron than it is outside. Typically, sodium ions (Na^+) are kept at a low concentration inside the neuron, whereas potassium ions (K^+) are kept higher inside. This is done by special molecular pumps in the cell's membrane. If a gate is opened that can pass both types of ions, the sodium ions will flow in and the potassium ions will flow out.*

When nothing much is happening, the neuron has a "resting" potential across its membrane. This is typically about -70 millivolts (inside versus outside). A change that makes this more positive at the

*This account is oversimplified, since the flow also depends on the potential difference across the membrane.

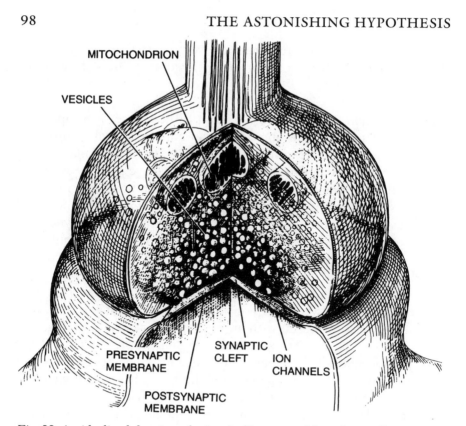

MITOCHONDRION

VESICLES

PRESYNAPTIC
MEMBRANE

SYNAPTIC
CLEFT

ION
CHANNELS

POSTSYNAPTIC
MEMBRANE

Fig. 32. An idealized drawing of a "typical" synapse. Note the small synaptic cleft.

cell body (say, -50 mV) is likely to make the cell fire. One that makes the potential even more negative can prevent it from firing at all. Whether the neuron is sufficiently excited to produce a spike in its axon depends on how much these potential changes (produced by activated synapses at different places on the dendrites and cell body of the neuron) alter the electrical potential at a special region near the beginning of the axon.

Let us look a little more closely at a synapse (see Fig. 33). There are two main types in the cortex, called Type 1 and Type 2, which can be distinguished in an electron micrograph.* Type 1 synapses usually excite and Type 2 usually inhibit the recipient neuron.

*Type 1 synapses have round synaptic vesicles, whereas those of Type 2 are often ellipsoidal or flattened. Type 2 synapses are more symmetrical than Type 1 and their synaptic cleft is usually a little smaller.

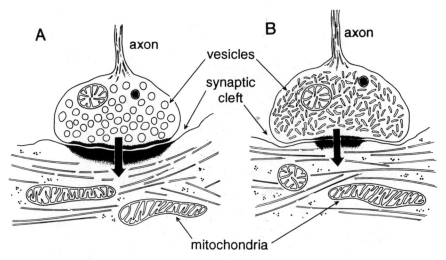

Fig. 33. The two main types of synapses found in the cortex: A, Type 1 (excitatory); B, Type 2 (inhibitory). In each figure the axon is at the top, the dendrite at the bottom, with the synaptic cleft in between. The arrows show the direction of the main flow of information, from the (presynaptic) axon to the (postsynaptic) dendrite.

   The main type of excitatory synapse in the brain does not lie directly on the shafts of the dendrites but on short little twigs attached to them, called "spines" (see Fig. 34). A single spine never has more than one Type 1 (excitatory) synapse, although some spines also have a single Type 2 (inhibitory) synapse. As can be seen from Figure 34, a spine is rather like a little flask attached by its neck to the dendrite. It has a roughly spherical head (usually somewhat distorted) and a thin cylindrical neck. The synapse itself lies on the head and is somewhat isolated from events elsewhere in the cell. It contains many receptors, including ion gates that can be opened if a molecule of neurotransmitter (from the synaptic cleft between the incoming axon and the recipient spine head) comes to sit on a special site on that kind of receptor molecule.
   A spine is a fairly elaborate structure and we are a long way from understanding its function completely. I suspect that a spine is a key evolutionary invention that allows for far more sophisticated processing of the incoming signal than would be possible without it.

I shall not attempt to describe the many types of protein molecules in the lipid membrane surrounding the neuron. Some of these are activated by

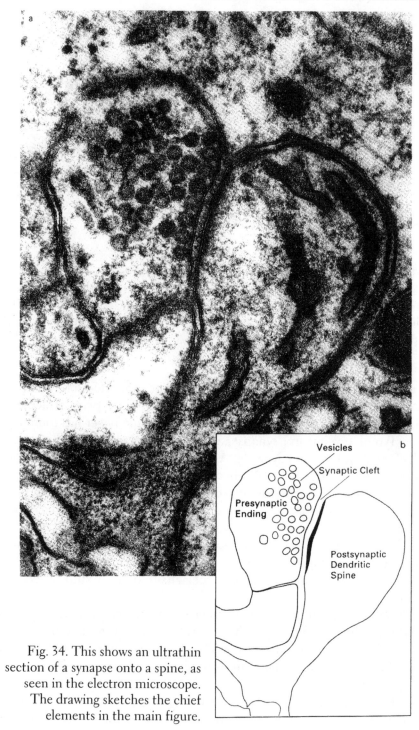

Vesicles

Synaptic Cleft

b

Presynaptic
Ending

Postsynaptic
Dendritic
Spine

Fig. 34. This shows an ultrathin section of a synapse onto a spine, as seen in the electron microscope. The drawing sketches the chief elements in the main figure.

transmitter molecules,* in which case they are called "receptors." The main excitatory transmitter molecule in our neocortex is a small, rather common, organic molecule called "glutamate."† Although there are two main classes of ion channels (one sensitive only to voltage and the other sensitive only to a neurotransmitter), a third, called the "NMDA channel," is the most interesting.‡ It is sensitive both to voltage *and* to glutamate. More precisely, it rarely opens, even in the presence of glutamate, if the local membrane potential is near its resting value. If this potential becomes less negative (due, say, to the activity of other excitatory synapses nearby on the dendrite), then glutamate will open the channel. Thus it only responds to an association between the presynaptic activity (due to the incoming axon releasing glutamate) and the postsynaptic activity (in the form of an alteration of the voltage across the local membrane produced by the inputs of others). This, as we shall see, is *a key property of brain function.*

When the NMDA glutamate channel opens, it allows the passage of not only sodium and potassium ions but also an appreciable amount of calcium ions (Ca^{++}). These incoming calcium ions appear to be the signal that initiates complex chains of chemical reactions that at the moment are only partly understood. The net result is that the strength of the synapse is altered, probably for days, weeks, months, or even longer. (This may form the basis for one particular form of memory—see the description of Hebb's rule in Chapter 13.) We can now begin to see the beginnings of the explanation of cognitive processes, such as memory, in terms of molecular events. An experimental example: If these NMDA channels are chemically inactivated in a rat's hippocampus, the rat cannot remember where it has been.

And what of inhibition? Is there a neuron whose axon produces excitation at some of its terminals and inhibition at others? Rather surprisingly, this never (or seldom) happens in the neocortex. More precisely, the terminals of the axon of a *particular* neuron are either all excitato-

---

*Some respond only to changes in voltage across the membrane. Others respond only when a particular small molecule—a neurotransmitter—sits on a part of the protein that lies *outside* the membrane. Some proteins have ion channels that can open rapidly to allow ions to flow through them; others have none. They produce slower effects inside the cell by indirect means, known rather cryptically as "second messengers."

†Glutamate is one of the twenty amino acids used to make proteins. It is sometimes used to add flavor to food.

‡A whole set of genes for this type of receptor has been isolated.[1]

ry or all inhibitory, never a mixture of both. The excitatory synapses appear to use glutamate as their neurotransmitter, as we have seen. The inhibitory ones employ a related small molecule called GABA.* In the neocortex about a fifth of all the neurons release GABA.†

An important consequence of the fact that most synaptic transmission is not electrical but chemical is that small, specially tailored molecules can interfere with it, often at very low concentrations. This is why the psychedelic effects of LSD can result from a dose as low as 150 *micro*grams. It also explains why certain drugs can, in some circumstances, alleviate mental conditions, such as depression, that appear to be caused by failures in the function of one form or another of neurotransmission. For example, the chemicals in sleeping pills (of the benzodiazepine family) are known to bind with the GABA receptor and make the effects of GABA greater. This increased synaptic inhibition promotes sleeping. The tranquilizers Librium and Valium are also benzodiazepines and act in a similar way.

Excitation and inhibition are not arranged symmetrically in the neocortex, although some theoretical models assume they are. The long-distance connections from one cortical area to another are provided only by pyramidal cells. These are all excitatory. The axons of most inhibitory neurons are rather short and only influence other neurons in the same neighborhood.‡ No type of neuron (with one possible minor exception) comes in two very similar morphological forms, one producing excitation and the other producing inhibition. The whole arrangement is asymmetrical in at least two ways: A neuron cannot send negative spikes, and quite different classes of neurons produce excitation and inhibition. However all neurons receive both excitation and inhibition, presumably to prevent them being always silent or from going wild.

There are only two main neurotransmitters in the neocortex: glutamate (or a near relative) for excitation and GABA for inhibition. Unfortunately, things are not that simple—there are a host of other

---

*There are two main types of GABA receptors. GABA Type A is a fast ion channel that is permeable to chloride ions (Cl⁻) whereas GABA Type B receptors act more slowly through a second messenger system.

†Such neurons, when mature, have few or no spines on their dendrites. Their synapses lie directly on their dendrites and on their cell bodies, and they typically fire faster than the excitatory, spiny neurons. There are several quite distinct types of inhibitory neurons, but it would take us too far afield to describe them all here.

‡There is one type, called a "basket cell," that does send inhibition over somewhat longer distances within a cortical area.

neurotransmitters. The neurons of the brain stem that project to the cortex use transmitters such as serotonin, norepinephrine, and dopamine. Other neurons in the brain use acetylcholine. About a fifth of the inhibitory cells also release somewhat larger organic molecules called "peptides," together with the usual GABA. Most of these transmitters produce slower effects than the two main fast transmitters (glutamate and GABA). They usually modulate the cell's firing rather than fire the neuron by themselves. Such transmitters probably mainly implement rather general processes, such as keeping the cortex awake, or telling it when to remember something, rather than dealing with the vast amount of intricate information handled by the fast processes.

Not only are there many neurotransmitters (even though only two of them do most of the work), but there are many kinds of channels. There are at least seven types of potassium channels, most of them fairly common.* Some open fast, others more slowly. Some channels, when opened, inactivate rapidly, others slowly. Some are mainly used to form the travelling axonal spike; others produce more subtle effects in the soma and dendrites. To calculate the exact behavior of a neuron to incoming signals we need to know the properties and distribution of all these channels in that particular neuron.

Different neurons fire in different ways. Some can fire very rapidly, others more slowly. Some fire single spikes, others tend to fire in bursts. In some cases, the same neuron can fire in either of these two modes, depending on its activation and on its recent behavior. Neurons may show a different type of firing in slow-wave sleep (the deep, dreamless form of sleep) than the type found in the awake animal, largely due to effects produced on the thalamus and the neocortex by neurons in the brain stem. Eventually we shall need a deeper and more complete understanding of the processes involved in neurons of all types.

A neuron, then, is tantalizingly simple when we look at it from outside. It responds to the many incoming electrical signals by sending out a stream of electrical impulses of its own. It is only when we try to figure out exactly how it responds, how this response changes with time, and how it varies with the state of other parts of the brain that we are overwhelmed by the inherent complexity of its behavior. Clearly, we need to understand just how all these chemical and elec-

---

*For example one of them, called Ic, is activated by *internal* concentrations of calcium ions (Ca^{++}).

trochemical processes interact. Then we need to boil down these responses so we can handle them, if only approximately, in a manageable form. In short, we need simple models for the different types of neurons, not so complicated that they are too difficult to follow and not so simple that we omit important features. This is more easily said than done. A single neuron may be rather dumb, but it is dumb in many subtle ways.

One characteristic of a neuron is already fairly clear. A single neuron can fire at different rates and, to some extent, in different styles. Even so, in any period of time it can only send out limited information. Yet during that time the potential information coming into it, through its many synapses, is very large. In this process—going from its input to its output—there must be a loss of information, at least if we look at one neuron in isolation. This loss is compensated by the fact that each neuron responds to particular combinations of its inputs and sends out this new form of information, not to just one place, but to many places: The pattern of spikes that a neuron transmits down its axon is distributed in much the same form to many different synapses because a single axon has many branches. What one neuron receives at one synapse is the same signal that many other neurons have also received. All this shows, if nothing else, that we cannot just consider one neuron at a time. It is the combined effect of many neurons that we have to consider.

It is important to realize that what one neuron tells another neuron is simply how much it is excited.* These signals will not normally give the receiving neurons other information—for example, where the first neuron is located.† The information in the signal will usually be related to certain activities in the outside world, such as that received by the photoreceptors of the eye.

In perception, what the brain learns is usually about the outside world or about other parts of the body. This is why what we see appears to be located outside us, although the neurons that do the seeing are inside the head. To many people this is a very strange idea. The "world" is outside their body yet, in another sense (what they know of it), it is entirely within their head. This is also true of your body. What you know of it is not attached to your head. It is inside your head.

---

*There may be some information in the pattern of firing, in addition to that coded by the average firing rate.

†A neuron can send chemical signals along its axon. These may in some cases convey additional information about the neuron, but the rate is too slow to convey fast information.

Of course, if we open the skull and pick up the signals sent out by a particular neuron, *we* can often tell where that neuron is located, but the brain we are studying does not have this information. This explains why we do not normally know exactly where our perceptions and thoughts are taking place in our heads. There are no neurons whose firing symbolizes such information.

Recall that Aristotle believed that such processes occurred in the heart because he could both locate the heart and observe its changes in behavior as a result of mental processes, such as falling in love. We cannot do this for the neurons in the human brain without the aid of special instruments. These, and others, are described in the next chapter.

9

Types of Experiment

"The art of research [is] the art of making difficult problems soluble by devising means of getting at them."
—Sir Peter Medawar

Strictly speaking, each individual is certain only that he himself is conscious. For example, I know I am conscious. Because your appearance and your behavior seem to me to be rather similar to mine, and in particular because you assure me that you are indeed conscious, I infer with a high degree of certainty that you, too, are conscious. It follows that if I am interested in the nature of my own consciousness I need not restrict my studies to experiments on myself. I can reasonably do experiments on other human beings, provided they are not obviously comatose.

If I wish to understand the neural basis of consciousness it is not enough to perform psychological experiments on alert people. I must also study the nerve cells and molecules in the human brain and the way they interact. Part of this information—mainly information about structure—can be obtained from the dead brain, but to study the complex activity of nerve cells we need to do experiments on the living human brain. There are no insurmountable technical problems about this, but overwhelming ethical considerations make many experiments of this type either impossible or very difficult.

Most people do not object to an experimenter fixing electrodes to their scalp in order to study their brain waves. They do, however, object to having a portion of their skull removed, even temporarily, so that electrodes can be stuck directly into the living brain tissues. Even if a person volunteered to have his head cut open—because he wished to further scientific discovery—no doctor would consent to perform the operation, saying either that it was against his Hippocratic oath or, more realistically, that somebody would be sure to sue him for doing it. In our society, you can volunteer for the armed forces and run the risk of being wounded or killed, but you may not volunteer to undergo dangerous experiments merely to obtain scientific knowledge.

A few brave researchers have done experiments on themselves— J. B. S. Haldane, the British biochemist and geneticist, was a striking example. He even wrote an article about it, entitled "On Being One's Own Rabbit." There have also been a few heroic medical episodes, such as Sir Ronald Ross showing that malaria is carried by mosquitos, but beyond that people are either discouraged or forbidden to volunteer for experiments that might help to satisfy scientific curiosity.

It is sometimes possible to do limited experiments on the exposed human brain of an alert person (with his informed consent) during certain necessary brain operations. There are no pain receptors in the brain, so the patient experiences no discomfort if the surface of his exposed brain is lightly stimulated electrically. Unfortunately, the time available for experiments during the operation is usually rather short and few neurosurgeons are sufficiently interested in the exact workings of the brain to attempt them. Such work was originally pioneered by the Canadian neurosurgeon Wilder Penfield in the middle of this century. In recent years, the leader in this field has been George Ojemann of the University of Washington School of Medicine at Seattle. Ojemann uses transient electric currents that are sufficiently small that they can inactivate a small local region near the electrode without producing any permanent after-effects when the current is turned off. He has concentrated on regions of the cortex that are involved with language, since when he removes portions of a patient's cerebral cortex, to relieve otherwise incurable epilepsy, he wants to trespass as little as possible on the adjacent language areas.

One of Ojemann's most striking results[1] was obtained from a patient bilingual in English and Greek. Stimulation at certain places on the surface of the left neocortex temporarily prevented her from using certain English words but not Greek ones. At other places the reverse was true, thus showing a clear difference in the location of some aspects of the two languages.

*   *   *

For most purposes, the activities of the human brain can be studied only from outside the skull.* There are various scanning methods that produce images of the living brain, but they all have serious limitations, either to their spatial or their temporal resolution. Most of these scans are rather expensive and are used strictly for medical reasons.

It is not surprising, therefore, that neuroscientists have often preferred to work on animals. I am less certain that a monkey is conscious than I am that you are, but I can reasonably assume that a monkey is not a total automaton—meaning a bit of machinery showing somewhat complex behavior but completely lacking in awareness. This is not to say that a monkey has the same degree of *self*-awareness as humans have. Experiments involving recognition in a mirror suggest that while some apes, such as chimpanzees, may have a degree of self-awareness, monkeys have little if any. But it seems a reasonable risk to assume that a monkey has a form of visual awareness not unlike our own even though he cannot express it in words. For example, the macaque monkey can be made to discriminate between two rather similar colors. These tests show the monkey's performance to be comparable to our own, within a factor of two or so. This is much less true for cats (they are mainly nocturnal) and even less so for rats. Few invasive experiments are done on the visual systems of chimps and gorillas, if only because they are too expensive. If we are mainly concerned with the molecules in the mammalian brain, the best and cheapest animal is a rat or a mouse, since their brain molecules are likely to be very similar to ours, even though in many other respects their brains are simpler.

Monkeys and other mammals have another advantage over humans: They are, at present, much better subjects for neuroanatomy. This is because almost all the modern methods of studying the longer connections within the brain make use of the active transport of molecules that goes both up and down the nerves. To do so a chemical is injected into one part of a living animal's brain and then is allowed to travel along the connections (usually for several days) to the parts of the brain directly joined to the injection site. Then the animal is killed in a painless manner and its brain examined to see where the injected chemicals have gone. Obviously, using humans for such experiments is

---

*In rare cases it is necessary, for medical reasons, to implant permanent electrodes deep in the brain tissue, but rather few electrodes are used so that only a very limited amount of information can be obtained in this way.

out of the question. Because of this limitation we understand far more about the details of the longer connections of the macaque brain than we do of our own.

One might expect such an obvious gap in knowledge would be a matter of grave concern to neuroscientists, and that they would be clamoring for new methods to study human neuroanatomy, since human brains will not be exactly the same as macaque brains. This is far from being the case.[2] It is certainly time for one or more of the more farsighted foundations to start a crash program to invent new techniques to remedy the present backwardness of human neuroanatomy.

Even if new methods are devised, so that much better neuroanatomy can be done on humans, there are still many key experiments that can only be performed on animals. Most of these experiments produce little if any pain, but when they are over (in some cases they may last for months) it is usually necessary to sacrifice the animal, again quite painlessly. The animal rights movement is surely correct in insisting that animals be treated humanely, and as a result of their efforts animals in laboratories are now looked after somewhat better than they were in the past. But it is sentimental to idealize animals. The life of an animal in the wild, whether carnivore or herbivore, is often brutal and short compared to its life in captivity. Nor is it reasonable to claim that since both animals and humans are "part of Nature" that they should be entitled to exactly equal treatment. Does a gorilla really deserve a university education? It demeans our unique human capabilities to insist that animals should be treated in precisely the same way as human beings. They should certainly be handled humanely, but it shows a distorted sense of values to put them on the same level as humans.

If monkeys can be useful experimental subjects for neuroanatomy and neurophysiology, what are their limitations? It is possible to teach alert monkeys to perform simple psychological tests, but the process is very laborious. It may take several weeks or more to train a macaque to hold fixation (to stare at the same spot) and to press one lever when he sees horizontal lines and another lever when he sees vertical ones. How much easier to get graduate students to do this. Moreover, humans can describe in words what they have just seen. They can also tell us what they are imagining or what they have just dreamt. It is almost impossible to get such information from a monkey.

Only one strategy seems possible. This is to do some types of experiments on humans and other types on monkeys. This involves making certain risky assumptions about how similar (and how different) monkeys' brains are to ours. Rapid progress is impossible without risks, so we must be both bold, in proceeding in this way, and cautious, by checking our assumptions as often as we get the chance.

The oldest method for studying brain waves—the electroencephalograph, or EEG—involves placing one or more large electrodes directly onto the scalp. There is plenty of electrical activity inside the brain, but the electrical properties of the skull act as something of a barrier to picking it up. A single electrode will respond to the electric fields produced by many tens of millions of nerve cells, so that the contribution of an individual cell is quite submerged by the activities of its many neighbors. It is rather like trying to study human conversation in a city from an altitude of a thousand feet. You would hear the roar of a crowd at a football game but might have trouble deciding what language was being spoken there.

The great advantage of the EEG is that its discrimination in time is rather good, in the range of a millisecond or so. Thus the rising and falling of the brain waves can be followed rather well. What is less clear is what the waves signify. They are clearly very different in an awake brain from the waves seen in slow-wave sleep. In REM (rapid eye movement) sleep the brain waves are very similar to those in an awake brain, hence its other name, paradoxical sleep, since the person is asleep but his brain appears to be awake. It is in this phase of sleep that most of our hallucinoid dreams occur.

A technique often used is to record the brain waves not just at any old time but immediately after some perceptual input, such as the sound of a sharp click in one ear. The response to the stimulus is usually very small compared to the electrical background signals (the signal-to-noise ratio is low) so that a single response shows very little. The event has to be repeated many times, and all the signals averaged, lined up from the beginning of each event. This improves the signal-to-noise ratio (since the noise tends to average out) and often provides a fairly reproducible trace of the typical brain waves associated with that brain activity. For example, one fairly common peak in the response is called the P300—P for Positive, 300 for the 300 milliseconds between the time the signal is given and the peak (see Fig. 35). It usually correlates with something that is surprising and demands attention. My suspicion is that it is largely a signal from the brain stem to the higher parts of the brain that the event should be remembered in some way.

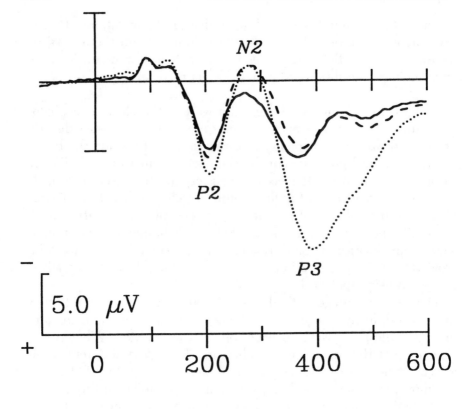

Fig. 35. Average evoked potentials showing the various components. The
P300 component is marked P3. The results are for three types of event: no
pop-out (solid), nontarget pop-out (dashed), and target pop-out (dotted).
Note the big P300 for the latter.

Unfortunately, there is a difficulty in locating the activity that produces these so-called event-related potentials (potential = voltage).
The mathematics of the problem shows that if one knows the electrical activity of every nerve cell, it is possible to calculate the effect on
an electrode at any chosen place on the scalp. Unfortunately there is
no way to do the reverse—that is, to calculate the electrical activity in
all parts of the brain from the activity in the electrodes. There are, in
theory, almost an infinite number of distributions of brain activity that
could produce the same scalp signal. Nevertheless, one would like to
have some idea of where most of the activity is taking place even if we

cannot recover all the details of it. We can get a better idea of where most of the activity is located by distributing a number of electrodes all over the scalp. If one electrode showed a big signal and all the others very small ones, then most of the activity is likely be near the active electrode. Unfortunately, in practice the situation is more complex than that.*

Some limited but useful information can be obtained from these event-related signals. For example, the auditory part of the cortex is mainly in the region of the brain near the temples. What goes on there when a person is born completely deaf? In one study the deaf subjects chosen were born of deaf parents, so that the underlying fault was almost certainly a genetic one, and probably in their ears rather than in their brain. By observing the event-related potentials, psychologist Helen Neville and her colleagues have shown[3] that some of the responses to visual signals in the periphery of the visual field have a much bigger peak (at a delay time of about 150 milliseconds) than do subjects who hear normally. This increase occurred in both the anterior temporal region (an area normally associated with audition) and in part of the frontal region of their brains.

It is not surprising that this increase is due to signals coming from the periphery of the visual field, since when deaf people sign to each other they mostly fix their gaze on the eyes and face of the signer. For this reason, much sign information comes from regions to the side of their center of gaze. As a control, Neville studied hearing subjects (of deaf parents) who had learned American Sign Language (ASL). They did not show the increase in activity displayed by those born completely deaf. This proves that it was not just the learning of ASL by itself that produced the effect.

Neville surmised that somehow parts of the visual system take over parts of the hearing system during the brain's development, since the normal sound-related activity will be absent there in these completely deaf people. In hearing people the normal auditory input presumably prevents any visual takeover of the auditory areas of the cortex. Recent experiments on animals make this idea plausible.[4]

---

*An approximation now being used is to assume that there are, say, four centers in the brain that are producing most of these electric signals. Then it is possible, by mathematical techniques, to locate the approximate positions of these centers. To test how good that assumption was one then assumes five centers and repeats the calculations. If it turns out that four of these are strong and one very weak, then the approximation of four is probably a fairly good one. Even so, it is really only an informed guess.

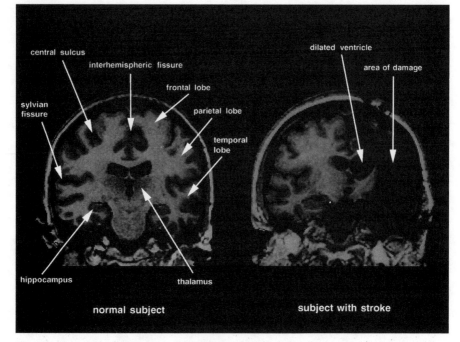

Fig. 36. A typical MRI scan showing the effects of a stroke.

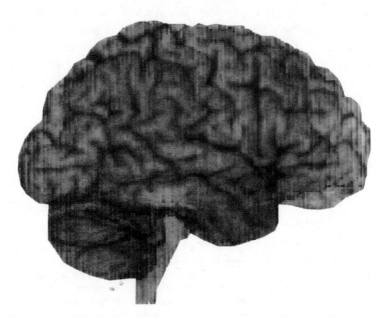

Fig. 37. A view of the living brain of neurophilosopher Patricia Churchland, synthesized from MRI scans by Hanna Damasio.

*   *   *

A more recent technique studies the varying magnetic fields produced by the brain. These are extremely small, only a minute fraction of the earth's magnetic field, so special detectors, called "squids" (for superconducting quantum interference devices) are used, and the whole setup has to be carefully shielded from varying magnetic fields in the environment. Originally, only one squid was employed, but now groups of thirty-seven are being used. This often produces better spatial localization than the EEG does. Otherwise the attractions and limitations are much like those for electric fields except that the skull interferes much less with magnetic signals. The magnetic detectors respond to sources (dipoles) perpendicular to the electric dipoles that affect the EEG, so the squids can pick up signals that the EEG misses, and vice versa.

Studying brain waves is not especially expensive, although the squid detectors are not cheap. The other main scanning methods not only require very costly apparatus but are also expensive to run. There are only a limited number of such scanners and they are almost all found in medical establishments. They all produce pictures of one slice of the brain at a time, so to cover any region of interest several slices are usually required.

There are broadly two kinds of scan: those that respond to some aspect of the static structure of the brain and those that detect activity. The oldest type, the CAT scan (Computer Assisted Tomography) uses X-rays. A more recent technique, which gives excellent high-resolution pictures, is the MRI (Magnetic Resonance Imaging). As far as can be told, it does no harm to the subject's brain. As normally used, it records the density of protons (the nuclei of hydrogen atoms) and thus is especially sensitive to water. It gives pictures with a good contrast, but the pictures are static and do not register the activity of the brain (see Fig. 36). Nevertheless, they clearly show the gross structural differences between one brain and another. Both methods can, in favorable circumstances, pick up structural damage to the brain (produced by strokes, bullet wounds, etc.), but some types of damage are more easily seen with one technique, some types with the other. By employing a special technique, MRI scans can be used to produce a 3D reconstruction of the brain of a living person, including views of its outside. Figure 37 shows one side of the (living) brain of the neurophilosopher Patricia Churchland.

*   *   *

A different method is PET scanning (Positron Emission Tomography). This can record local activity in the brain but only averaged over a minute or so. The subject is injected with a chemical molecule (often H_2O) that has been tagged with a harmless radioactive atom such as ^{15}O that emits a positron* when it decays. This tagged water goes into the blood. The short half-life of the ^{15}O means that the time from its production in a cyclotron to its injection must be kept very short, but it has two advantages. The oxygen decays so rapidly that within ten minutes or so a second experiment can be done; the short life of the radioactivity means that the total dose (for a required signal) to the subject is so small that its damaging effects are negligible. Thus it can be used on healthy volunteers rather than being restricted to sick patients.

When any part of the brain is more active than usual its blood flow increases. The map produced by the computer corresponds, in effect, to the level of blood flow to each brain region in the scan. A scan is also taken with the subject in some control state. The difference between the two maps corresponds roughly to the change in brain activity between the stimulated state and the control state.

This technique, especially in the hands of a group headed by Marcus Raichle at the Washington University School of Medicine in St. Louis, has produced numerous interesting and challenging results. In early experiments they studied the response of subjects to a small set of crude visual patterns, chosen to provide maximum excitation to different rather broad areas of the visual field. The changes of blood flow in the first visual area of the neocortex occurred approximately where they had been expected to be, based on knowledge gained earlier by studying the effects of brain damage in humans. Changes were also seen in other visual areas of the cortex but the results were not clear enough to be of much value.

More recently[5] they studied the changes in blood flow during a more sophisticated visual task, the so-called Stroop interference effect. In this experiment the subject has to identify as quickly as possible the color of a printed word. The catch is that the word *red* may be printed in, say, green. The disagreement between the color of the word (green) and the meaning of the word (red) increases the subject's reaction

*The positron wanders a short distance until it combines with an electron. Both particles are annihilated, their mass being turned into radiation in the form of two gamma rays travelling in almost exactly opposite directions. These gamma rays are recorded by a ring of coincidence counters. A computer combines the evidence from all the decays and works out the most likely regions of origin of the gamma rays.

time. The blood flow in such a task was compared with straightforward cases in which the word *red* was printed in red. In the Stroop condition they found an increased blood flow in several cortical areas but the largest was in a region called the "right anterior cingulate" (in the middle of the brain, near the front). They attributed this to the amount of attention required to accomplish the task. Their conclusion reads: "These data suggest that the anterior cingulate is involved in the selection process between competing processing alternatives on the basis of some preexisting internal, conscious plans." This sounds to me closer to what we think of as Free Will than it does to what we normally call attention (see the Postscript at the end of the book). Clearly, more needs to be known about the neuronal details of the various processes involved.

PET scans give us results that are difficult to obtain in any other way, but they have several limitations. Apart from the expense, the spatial resolution is not very good, although it is improving as more modern machines are developed. At present it is usually near 8 millimeters. The other disadvantage is the very poor time resolution—an appreciable fraction of a minute is required to obtain a good signal, whereas the EEG works in the millisecond range.

Leading centers are now combining a PET scan, for brain activity, with an MRI scan showing brain structure so that the results of the PET scan can be mapped onto that individual brain, rather than onto an "average" brain, as was done in the past. Before long, however, the interpretation of such results will run into the limitations, referred to earlier, of our lack of detailed knowledge of human neuroanatomy.

New methods of using MRI scans are also being developed. In one of these the instrument is adjusted so that it is preferentially sensitive to lipid.[6] The resulting pictures can help to locate some of the different cortical areas in that particular individual (their exact location may vary somewhat from one person to another). This is possible because some cortical areas have more myelinated axons (and therefore more lipid) than others.

Other new MRI methods attempt to detect various metabolic and other activities in the brain (rather than just its static structure) but some of these are likely to have a poorer signal-to-noise ratio than conventional MRIs. It will be interesting to see how these new methods develop. Several of them look promising.

So much for the study of human brains. What methods can be used to see how the neurons behave in an animal's brain? The technique that has produced the most detailed information uses a fine electrode. This is an insulated wire with a tiny exposed tip. It is placed right inside the nervous tissue after a section of the skull has been removed under an anesthetic. The electrode does not cause the animal pain since there are no pain detectors in the brain. Such a microelectrode can detect, from outside the nerve cell, when a nerve cell fires, provided the tip of the electrode is very close to that cell. It may also pick up weaker signals from more distant cells. By moving the tip of the electrode along its length, from one place to another in the tissue, it can listen to one nerve cell after another. The experimenter can select where he puts his electrode in the animal's brain, but exactly which type of nerve cell he records from is somewhat a matter of chance. Nowadays sets of electrodes are often used, so that one can listen to more than one nerve cell at a time.

Another technique is to study a thin slice of nervous tissue taken from the animal's brain. In such cases the electrode is a very thin tapered glass tube. This form of electrode can, with care, be located so that its tip is actually inside a nerve cell. This can give more detailed information about the electrical activity within that nerve cell. (This technique can also be used on the intact brains of an anesthetized animal, but it is difficult in alert animals.) A brain slice in a suitable bath can be made to last for many hours. It can easily be perfused with different chemicals to see what effects they have on the behavior of the nerve cell.

Neurons taken from the brains of a very young animal can, in some cases, be grown spread out on a dish. Such a neuron, as it grows, may make contact with a neighboring nerve cell. This condition is even further from the living animal, but it can be used to study the basic behavior of neural interconnections. Such connections have channels in their membranes that, when they open, allow charged atoms (= ions) to flow through them.

What perhaps is really remarkable is that it is now possible to study the behavior of a single molecule of an ion channel. This is done by a technique[7] known as "patch-clamping." For developing and exploiting this technique, Erwin Neher and Bert Sakmann received a Nobel Prize in 1991. A tiny glass pipette, with a specially bevelled tip of small diameter (say, 1 or 2 micrometers) can be manipulated to pick up a small piece of lipid membrane. With luck, this will contain at least one ion channel. By passing its current through an electrical amplifier and a recording device, its electric flow can be studied. The concentration of the relevant ions is maintained at different values on the two sides of the little patch of membrane. When the channel opens, even briefly, a large number of electrically charged ions rush through it. This stampede produces a measurable current, even though only a single channel is present. Thus the effects of neurotransmitters and other pharmacological agents—usually other small organic molecules—can be studied, as well as the effects of membrane voltage.

Patch-clamping can also be used to study ion channels whose genes have been artificially introduced into an unfertilized frog's egg. Following the instructions of these foreign genes, the oocyte (the unfertilized egg) will synthesize the proteins of the channel and deposit them in its external membrane where they can be picked up by the little pipette used in patch-clamping. This technique is helpful when trying to discover the gene for a particular ion channel.

In summary, there are many methods for studying the brains of men and animals, some from outside the head, others by going inside it. All have limitations of one sort or another, either in time resolution, space resolution, or expense. Some can be interpreted fairly easily but provide only limited information. Other measurements are easy to do but difficult to interpret. Only by using a combination of methods can we hope to unravel the mysteries of the brain.

10

The Primate Visual System—
Initial Stages

"I spy, with my little eye, something beginning with . . ."
—Children's game

Seeing is a complicated process, so it is not surprising that the
visual parts of the brain are not simple. They consist of one very
large primary system, one secondary system, and a number of
minor systems. All receive their input from some of the million or so
neurons, the so-called ganglion cells, at the back of each eye. The primary system connects to the neocortex via a small part of the thalamus called the Lateral Geniculate Nucleus (LGN). The secondary
system projects to the superior colliculus, mentioned earlier.

The general structure of the eye is well known (Fig. 38). It has a lens,
whose focal length can vary—at least in people under forty-five. The
aperture, called the "pupil," can also change, being smaller in brighter
light. The lens focuses the image of the visual field onto a thin sheet
of cells, the retina, at the back of the eye. In one of its layers are located the four kinds of photoreceptors that respond to the photons of the
incoming light. These are the rods and the three types of cones, so
named because of their shape. The rods, of which there are over 100
million in each eye, respond mainly in dim light and are of only one

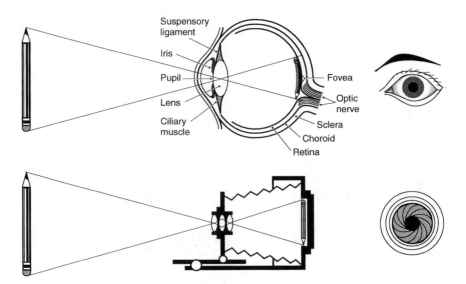

Fig. 38. The structure of the eye, with a camera for comparison.

type. The cones, of which there are about 7 million, are active in bright light. There are three types of them, each type responding to a different range of wavelength of the incoming light. Because of this we are able to see color, as already explained in Chapter 4.

Before passing on this incoming information, the retina first processes it. In fact, the retina is a little bit of the brain. Compared to the neocortex, it is relatively easy to study. The American physiologist John Dowling has called it "an approachable part of the brain." It will probably be the first part of the vertebrate brain that will be understood fairly completely. Interesting though its structure may be, I shall treat it as a "black box" and merely describe how its output (the firing of its ganglion cells) relates to its input (the light falling on the eye).*

The density of the cones used for daytime vision is very much greater in the fovea—approximately at the center of the eye—and so we can see much finer detail there. This is why you switch your gaze to something of interest in order to see it more clearly. Conversely, you can sometimes see in the dark more clearly out of the corner of your eye, where the retina has many rods.

The eye can move in different ways. It can make jumps, called "sac-

---

*In mammals, very few neurons, if any, project from the rest of the brain to the retina, although of course we can influence what happens there by moving our eyes.

cades," usually three or four times a second. The eyes of primates can
follow a moving object, a process called "smooth pursuit." Curiously,
it is almost impossible to move your eyes smoothly over a stationary
scene by just willing to do so. If you try to, it will move in jumps. The
eye also makes continual tiny movements of various sorts. If, by one
means or another, the image on the retina is held completely stationary, it fades from consciousness after a second or two. (This is discussed more fully in Chapter 15.)

The cells that send signals from the eye to the brain are called "ganglion cells." Any particular ganglion cell will only respond vigorously to
a small spot of light turned on (or off) in *one particular part* of the
visual field (Fig. 39). The spot has to be in that place because the lens
focuses the spot to a position near that of the ganglion cell in the retina. This will depend on where the eye is pointing. (In the same way
the response of a particular tiny part of a photographic film in a camera is related to its position on the film and to the direction in which
the camera is pointing.) The part of the visual field that influences a
particular cell is called the cell's "receptive field."
    In total darkness, a ganglion cell usually fires at a low, irregular rate,
called its "background rate." For one type of ganglion cell—the so-
called on-center type—its firing will increase dramatically if a small
spot of light is shone onto the very center of its receptive field. This
small center has a circular region surrounding it where exactly the

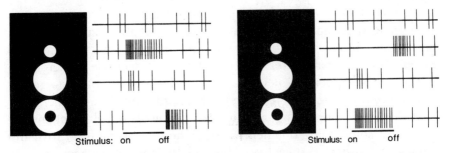

Fig. 39. Recordings from typical ganglion cells. Those on the left are of the
"on-center" type; those on the right are "off-center." Each short vertical line
represents an axonal spike. The stimuli are shown within the two black rectangles. The top traces show the background rate of firing, when no light falls
on that part of the retina. The lower three show responses to a small spot of
light, a large spot of light, and a ring of light with a dark center.

opposite happens. If the spot falls entirely in this annular region, the background rate of firing ceases. If the spot is then turned off, there is a burst of rapid firing (see the left side of Fig. 39).

Suppose we position spots of light of various diameters, all centered on the very middle of the cell's receptive field. As we have seen, a small spot, when turned on, makes the cell fire vigorously, but for spots of larger diameter the response is less, and when the spot is large enough to fill both the center and the surrounding annulus there is hardly any response at all. In other words, the response of the center of the receptive field is counteracted by its surround. This means that any particular ganglion cell will respond with rapid firing to a small spot of light in just the right position but will hardly respond at all to a patch of uniform illumination in that general region. The retina is processing the information coming into the eye in such a way that it partially eliminates redundant information. What is sent to the brain shows mainly the interesting parts of the visual field, where the light distribution is not uniform, largely ignoring the dull parts where it is fairly constant.

In addition to on-center cells, there are an equal number of off-center cells. They are, loosely speaking, the opposite of the first type—that is, the cells fire rapidly when a spot in the middle of their receptive field is turned off (see the right side of Fig. 39). This illustrates a rather general property of many neurons that send spikes down their axons. A neuron can't produce negative spikes. How then can a negative signal be sent? In the thalamus or the cortex it is unusual to find a fast background rate of firing of, say, 200 Hertz. Such a cell, if it existed, could signal a positive response by increasing its rate of firing up to 400 Hertz and signal a negative response by lowering its firing rate toward zero. Instead of such a cell there are often two, rather similar types of neurons, each with a low background rate of firing. One will fire to an increase in some parameter and the other to its decrease. When nothing much is happening, neurons usually do next to nothing—rather than buzz away at 200 Hertz—probably in order to conserve energy.

If the brain wants to signal a simple sinusoidal (wavelike) change of activity at some point, one neuron fires when the signal is positive and another one when it's negative. This should caution one against using mathematical functions in too simple a way to describe what is happening. Moreover, a real neuron often responds to a sudden change in its input with an initial burst of firing, usually of somewhat limited duration; the exact temporal firing pattern can vary from one type of neuron to another. Neurons were not evolved for the convenience of mathematicians.

The size of a ganglion's receptive field (that is, the visual solid angle to which it is sensitive) varies considerably, being much smaller near the center of the eye than it is near the periphery. Ganglion cells are always relatively close together, so that the receptive fields of adjacent cells overlap somewhat. A spot of light on the retina will usually excite a whole group of adjacent ganglion cells, though not necessarily all to the same degree.

There are not just two main types of ganglion cells, on-center and off-center. There are really several broad *classes* of ganglion cells (each having on- and off-center subtypes). The exact nature of these classes in mammals varies somewhat from species to species. For the macaque monkey, there are two principal classes,* sometimes called "M cells" and "P cells" (M stands for magno, meaning large; P for parvo, meaning small). Human ganglion cells are probably fairly similar. In any part of the retina the M cells are larger than the P cells, and have larger receptive fields. They have a thicker axon and this makes their signal travel to the brain faster. They respond well to small differences in light intensity and so handle low contrast well, although their firing rate reaches a plateau at high contrast. They are probably used largely to signal changes in the visual scene.

The P cells are more numerous. Their responses are more linear (proportional to the input) than most M cells. They are more interested in finer detail, higher contrast, and especially color. For example, the center of the receptive field of a P cell may respond well to green wavelengths whereas its antagonist surround may be more sensitive to red ones. Because of this there are several subtypes of P cells, each interested in different color contrasts. We see, once again, that the retina is not just transmitting raw information about the light falling on its photoreceptors. It has started the job of processing this information and is *doing so in more than one way*.

The two main classes of ganglion cells, the M and the P cells (each having on-center and off-center members), send their axons to the LGN, the region of the thalamus that relays the information to the neocortex. The retina, however, also projects to the superior colliculus. P cells do not project in this way, although some M cells do, as do a number of the minor, miscellaneous cell types. The lack of P-cell input means that the colliculus is colorblind.

---

*There is a third, rather numerous class, sometimes called "W cells," with somewhat miscellaneous properties.

*   *   *

In most vertebrates the ganglion cells of the right eye project almost entirely to the optical tectum (roughly the equivalent of the superior colliculus in mammals) on the *left* side of the brain and vice versa. In primates, matters are more complicated. Each eye projects to both sides of the brain, but it does so in such a way that the left side of the brain receives input relating only to the right half of the *visual field*.

Thus everything you see to the right of your center of gaze goes to the left LGN, on its way to the left visual cortex (Fig. 40) and also to the left superior colliculus. Of course, the two halves of the brain are normally connected to each other by several tracts of nerve fibers of which the largest is the corpus callosum. If this is cut (for medical reasons), as we shall discuss in Chapter 12, the left half of the brain of that person only sees the right side of the visual field and the right half only the left side. This can produce somewhat surprising results, almost as if there were now two persons in one head.

Let's first consider briefly the secondary system that projects to the superior colliculus. This is the main visual system in lower vertebrates, such as the toad; in mammals, many of its functions have been taken over by the neocortex. Its main remaining function appears to be the

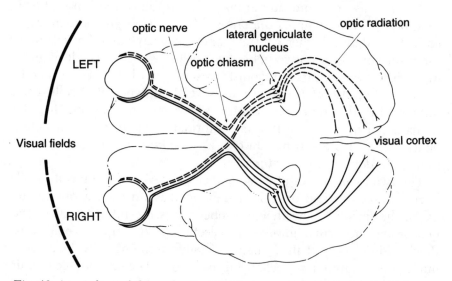

Fig. 40. An outline of the early visual pathways seen from underneath the brain. Notice that the right visual field influences the left side of the brain, and vice versa. The connections related to the right visual field are shown by lines of dashes.

control of eye movements, but it is probably also involved in other aspects of visual attention.

The colliculus is a layered structure. There are three main regions—I shall call them the upper, middle, and deep regions. The upper regions receive various kinds of retinal input as well as input from the auditory and somatosensory systems. The inputs are crudely mapped, although the details of the mapping vary from species to species. The inputs to the deep regions are more various.

It is important to know that in the deep region there are neurons that connect with the colliculus on the other side of the brain, via a pathway called the "intertectal commissure." (This pathway is usually left intact in the split-brain operations described later in Chapter 12.) These deep regions also connect to the brain stem onto neurons that lead to the control of the muscles of the eyes or the neck.

What about the behavior of the neurons. In the upper region many of them are selective for movement. In the macaque they are color-blind—that is, they do not respond selectively to the wavelength of the light. They are very interested in small stimuli, but less in the details of a stimulus. Their response to a change of light, either on or off, is often very transient. These are all factors that are likely to command involuntary attention. They signal "look out, there's something there!"

Anyone who lectures is likely to have had the following experience. If there is a sudden change, such as a door opening, to the left or the right of the speaker, the eyes of the entire audience swivel simultaneously in that direction. The response appears to be almost immediate and largely involuntary. I would certainly expect that their colliculi are a major factor in producing such eye movements.

But how do the eyes know where to jump? Thanks to very elegant experiments by David Sparks, David Robinson, and others, we now have a much better idea of how this happens.[1] Whereas the upper region of the colliculus might be described as a sensory map, the middle and deep regions appear to embody a "motor map." The firing of the neurons in these regions encode the direction and amplitude of the change in eye position needed for the eyes to make a saccade to the target. This signal is more or less independent of the position of the eyes at the moment before the jump begins. The message it sends to the brain stem is how big a jump to make and in what direction.

This signal is not expressed in the way an engineer might have guessed. For example, one neuron might have coded one particular direction for the jump and its rate of firing might have coded for the distance of the jump. In this way a small set of neurons could code for all directions and all distances. Alternatively (but using rather more neurons) each separate neuron might code for just one particular jump vector—that is, the direction and the distance of the jump. The truth is quite different. To produce a saccade, a patch of collicular neurons starts to fire rapidly. Broadly speaking, it is the center of this activity in the motor map that determines the jump vector. Thus a particular single collicular neuron may take part in many rather different kinds of jumps. It is the active neurons as a whole that determine the vector nature of the saccade. In short, the control of a single eye movement is controlled by many neurons.*

And what controls the speed of an eye movement? It may be correlated with the rate of firing of the neurons of the active patch. The faster they fire, the faster the eyes move. Thus the final direction of the jump does not depend on exactly how fast the relevant neurons are firing but only on the position of the effective center of the active group in the motor map.

You may find this arrangement very peculiar but it is a beautiful (and typical) example of how a set of neurons can code related parameters, such as the speed and direction of eye movements. Its advantage is that the system will not fail if a few neurons became inactive. No engineer would design a system like this unless he had already learned how the brain does it. When these signals arrive at the brain stem they have to be transformed into a different set of signals to control the muscles of the eye. Exactly how this is done has yet to be discovered.

Let's now consider the primary visual system that projects through the LGN to the visual cortex. The LGN is a small part of the thalamus. When I went to the Salk Institute in 1976, I inherited the office, overlooking the ocean, that had belonged to the late Bruno Bronowski (who made the TV series "The Ascent of Man"), together with a large, colored plastic model of the human brain, twice life size. One of the first things I did was to try to locate the LGN on it. I easily found the thalamus, but it took me some time to locate the small bump on it that denoted the LGN. This is not surprising as it consists of only about a million and a half neurons.

There are two things to grasp about the LGN. The first is that it looks like a relay and nothing more. The second, in contradiction to

---

*Note, however, that the required output is only a simple two-component vector, so this method cannot be used when an area has to handle much complicated information at the same moment.

the first, is that it is probably doing something a lot more complicated than this which we do not yet fully understand.

The main neurons in the LGN—the principal cells—produce excitation. (In addition, there are a minority of GABAergic cells that produce inhibition.) The LGN is called a relay for two reasons, one is anatomical, the other physiological. The principal cells receive input directly from the retina and send their axons directly to the first visual area (V1) of the neocortex. There are no other neurons in between in this pathway—hence the name "relay." These axons have very few collateral branches to other principal cells or to the other parts of the LGN. In other words, these neurons tend to keep to themselves and not talk much to their fellows. In addition, the retinal input is mapped onto the LGN so that each layer of the LGN has a somewhat distorted map of the visual field. The receptive fields of these LGN neurons are not unlike those of the retina, although they are sometimes a little larger. At first sight the LGN merely passes the retinal information to the visual cortex in more or less the same form as it received it.

The word *map* is used in two somewhat distinct ways in the visual system. The general meaning of mapping is that axon terminals that end near each other in the recipient area usually originate from neurons that are not too far apart in the donor area. This necessarily produces a rough sort of map of the donor area in the recipient one. The more restricted meaning is usually called "retinotopic" mapping. This means that neurons near each other in that particular visual area tend to respond to the activity of points near each other on the retina (and thus near each other in the 2D projection of the 3D visual field). As one proceeds farther into the visual system, the retinotopic mapping gets increasingly jumbled (due to the many stages of approximate mapping), but the neural mapping from one area to the next may still be fairly well preserved.

The macaque LGN has six layers (see Fig. 41). Two of these have large cells (called "magnocellular"). One of them gets its input from the right eye, the other from the left eye. There is little interaction between the layers. Their input is mainly from the M cells of the retina. It might be thought that the P cells of the retina projected, in a similar way, to two layers with small cells (called parvocellular), but, just to make things more complicated, there are four parvocellular layers, not just two. Again the inputs from the two eyes are segregated. The main point is that the M and P inputs are largely kept apart, as are the inputs from the two eyes.

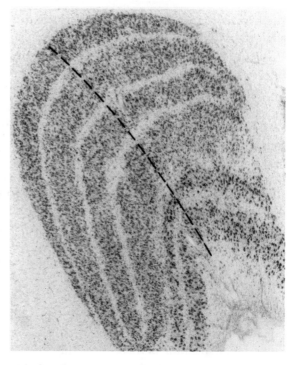

Fig. 41. The six layers of the LGN of a macaque monkey. The section is stained to show the cell bodies, each of which appears as a dot. The lowest pair of layers has larger (M) cells, and are called the "magnocellular layers." The upper four layers have smaller (P) cells and are called "parvocellular." Each layer gets input from only one eye.

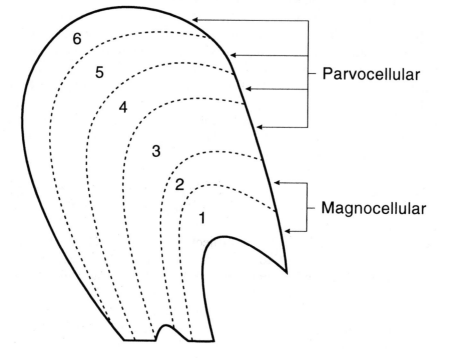

What is the difference between the roles played by the parvocellular layers and the magnocellular layers? Two laboratories have trained alert monkeys to perform various visual tasks and have then made small, local lesions in the LGN. These experiments suggest that, in the main, the neurons of the parvocellular layers carry the signals that relate to color, texture, shape, and stereopsis, and that the magnocellular neurons specialize in detecting movement and flicker (for a full review see reference 2).

So far we have only talked about the principal, excitatory cells. The inhibitory neurons fall into two main classes, those in the LGN proper and those in a thin sheet of cells called the "reticular nucleus of the thalamus" (not to be confused with the reticular formation in the brain stem). This thin sheet of cells surrounds much of the thalamus. Its neurons are *all* inhibitory. They receive excitation from most of the axons passing to and from the neocortex and they interact with each other. Their output is mapped onto the underlying part of the thalamus immediately beneath them. If the thalamus can be described as the gateway to the cortex, the reticular nucleus looks like the guardian of the gateway.

The LGN neurons also get input coming back from the first visual area (V1) of the cortex. Surprisingly, there are many more axons coming back from V1 than going up to it, but they tend to synapse onto those parts of the dendrites rather distant from the cell bodies of the LGN neurons, so their effects may be rather subdued. The exact function of these reverse connections is not known (see Chapter 16 for some speculative suggestions about them).

There are also inputs from the brain stem that modulate the behavior of the thalamus and especially its reticular nucleus. This means that the LGN freely transmits visual information in the awake animal but blocks this transmission somewhat when the animal is in slow wave sleep. There are many more details about the neurons in the thalamus and about their various types of synaptic connections, but what has been described about the LGN should convey its baffling combination of apparent simplicity with important complexities.

The principal cells of the LGN project to the visual cortex (see Fig. 40). The axons in cats go to several visual areas, but in the macaque monkey and in man they connect almost entirely to the first visual area.* (There is now believed to be some very weak connections to other visual areas of the monkey cortex. These will be relevant when

---

*Also called the "striate cortex" and "area 17."

we discuss "blindsight" in Chapter 12.) If a person or a monkey has extensive damage to all parts of V1, he is almost totally blind for that half of the visual field.

At first sight, any part of the cerebral cortex looks a complete mess, with about 100,000 neurons under each square millimeter. The axons and dendrites appear jumbled together, mixed up with numerous supporting glial cells and blood vessels, in a totally chaotic arrangement. They have none of the controlled orderliness of the transistors and other structures in a computer chip. On further scrutiny some partial degree of order begins to emerge. As the general arrangement of neurons in the many different parts of the cerebral cortex is much the same, let's look first at what many cortical areas have in common.

The cerebral cortex is a thin sheet—that is, its dimension perpendicular to the sheet is much smaller than its dimensions parallel to the surface of the sheet. The arrangement and appearance of the neurons is asymmetric. The direction perpendicular to the sheet is referred to as the "vertical" direction (as if the cortex had been unfolded and spread out on top of a table) and the other two directions as "horizontal." For example, almost all pyramidal cells have an "apical" dendrite that we say ascends "vertically" toward the pial (outer) surface of the cortex. By contrast, the two horizontal dimensions of the cortex are, on average, fairly similar to each other in their properties. This is rather like the arrangement of trees in a forest: The vertical direction appears very different from the two horizontal ones.

The most noticeable feature of the cortical sheet is that it is layered. It is important to know something about these layers as the neurons in the different layers do somewhat different things. It is conventional to describe six layers, but in fact there are often several sublayers within a layer (see Fig. 42). The top layer—layer 1—has few neuronal cell bodies. It consists mainly of the apical dendrites of many of the pyramidal cells situated in the lower layers together with various types of axons that make synapses on them. It is all wiring and few cell bodies. Below lie layers 2 and 3, often referred to collectively as the upper layers. These layers have many pyramidal cells. Layer 4 has many spiny (excitatory) stellate cells and very few pyramidal cell bodies. Its thickness varies very much from cortical area to cortical area, being almost absent in some of them. Layers 5 and 6, often referred to as the lower layers, again have many pyramidal cells. The apical dendrites of some of them reach all the way up to layer 1.

It is not only that the neurons in different layers are different, one

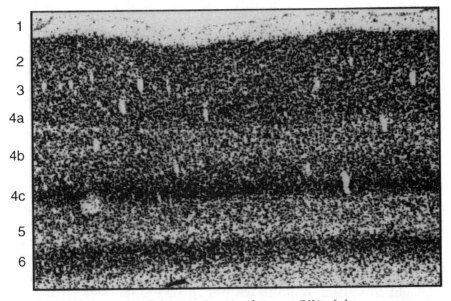

1
2
3
4a
4b
4c
5
6

Fig. 42. A cross-section of the primary visual cortex (V1) of the macaque monkey. As before, each dot represents the body of a single cell. Notice the layered structure. The numbering of the layers is shown on the left. (The white gaps are sectioned blood vessels.)

from another, but more importantly, their neurons are *connected* in somewhat different ways (see Fig. 43).

The upper layers (layers 2 and 3) talk only to other cortical areas. They never project out of the cortex as a whole, although some of their neurons may connect to cortical areas on the other side of the head, via the corpus callosum. Some of the neurons of layer 6 project back to the thalamus or to the claustrum, a thin satellite of the cortex lying just under it, toward the middle of the brain, although some of them have axon collaterals that connect to neurons in layer 4. The only part of the cortex whose neurons project completely out of the cortical system—that is, not to other parts of the cortex, nor to the thalamus, nor to the claustrum—lie in layer 5, although some neurons there project to other cortical areas. In a sense, therefore, layer 5 is where the processed information in the cortex exits to other parts of the brain and to the spinal cord. All these connections that leave the cortical sheet, even if they then enter it again, are excitatory.

Of course the cortex also contains many inhibitory cells. Numerically, the pyramidal cells (which produce excitation) are in the majority. The inhibitory neurons, which use GABA as a neurotransmitter, usually make up about a fifth of the total, the rest being mainly spiney stellate cells. The axons of the spiney stellate cells (which produce excitation) are fairly short and only connect to neurons fairly near to them in the horizontal directions, say within 100 or 200 microns. This is also true for all but one of the several types of inhibitory neurons.*

There is one type of inhibitory cell that appears not to exist. The axon of a pyramidal cell usually travels downwards and leaves that region of the cortex for another often fairly distant destination. Before it does so it usually throws off several branches, called "collaterals." In some cases these form many branches locally, but they can also travel fairly long distances (as much as several millimeters) horizontally within that piece of cortex.

If we think of the cortex as carrying out computations, it might be thought that there would be a special type of inhibitory synapse—a sort of gate—that would allow the information leaving the cell body through its axon to circulate within that cortical region several times (so that it could perform computations demanding several reiterations) before sending the "result" out along the main branch of the axon to its destination in other areas. For this we would need a strong set of inhibitory synapses not at the start of the axon (where the synapses from the chandelier cells are found) but just before the axon exits from that part of the cortical sheet. Of such synapses there is no evidence, although at least one theorist has invented them in order to make his model work! Nor are there any at other branch points of the axon. All this suggests that a cortical area is in such a hurry to send out its messages that it performs hardly any iterations before it starts to do so. It probably implies that connections between various cortical areas can be as important as those within each single cortical area when the brain needs to establish a working coalition of activities by reiterative computation.

---

*The exception is a type of inhibitory neuron called a "basket cell." Its axons can travel much greater distances horizontally within the cortex, up to a millimeter or more. When they connect with another neuron they form multiple synapses on its soma (the cell body) and the proximal (nearby) parts of its dendrites. They therefore produce fairly powerful inhibitions in a crucial region of the neuron. Exactly what purpose they serve is unknown. We are also ignorant of the exact function of another remarkable type of inhibitory cell, the so-called chandelier cell. Its axons contact only pyramidal cells and then only at the beginning part of their axons, where it has multiple inhibitory synapses.

*   *   *

Can one say anything at all about the broad way the information flows between the various layers of the cortex? This is an enormously complex process, but an outline scheme would probably go as follows (see Fig. 43).

The main, but not the only, entrance to a cortical area is into layer 4 or, when this is small or absent, to the lower part of layer 3. Layer 4 connects mainly to the upper layers, 2 and 3, and they in their turn have a large local connection to layer 5. Layer 5 sends longish "horizontal" local connections to layer 6 beneath it, which in turn sends some short "vertical" connections back to layer 4. There are also important inputs (from other cortical areas) into layer 1. These can contact the apical dendrites of tall pyramidal cells from most of the lower layers.

This simple summary conceals the intricate nature of many of the axonal connections within a small piece of cortex, in particular the many connections of one layer to itself, some of them surprisingly long. There is clearly some logic behind all these regularities, but until

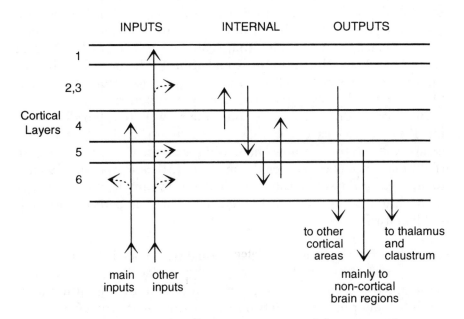

Fig. 43. A grossly simplified diagram showing some of the major pathways within cortical area V1. There are many sideways connections that are not shown in this diagram.

we understand the cortex better it is difficult to say exactly what it is. The neocortex may be the crowning glory of humans but it will not yield up its secrets easily.

One final point about the different regions of the cerebral cortex is needed. The cortex was originally divided up into regions based on the appearance of thin, stained sections of it under high-powered optical microscopes—such studies are called "architectonics." For example, the first visual area, V1, was named the striate cortex because it has marked horizontal striations produced by large tracts of axons running horizontally within it in all directions. These striations are large enough to be seen on a stained microscope slide with the naked eye (Fig. 44). The striations end abruptly at the edges of a large patch of cortex, so it was natural to regard this patch as a fairly uniform area of some sort and to give it a name or a number. Other patches of the cortex had somewhat different appearances. For example, the striate cortex has a very thick layer 4, whereas the primary motor cortex has little if any. Unfortunately, some of the distinctions between adjacent regions were so subtle that neuroanatomists did not always agree about them. At the beginning of the twentieth century a German neuroanatomist named Korbinian Brodmann described his division into distinct regions of the cortex of various mammals, including man, and gave each area a number. The striate cortex he called area 17, the area adjacent to it area 18, and the area adjacent to that area 19. The primary motor cortex he labelled area 4. Other neuroanatomists, such as Oskar and Cécile Vogt, discerned many more subdivisions.*

Brodmann's divisions have stood up fairly well, but, broadly speaking, they are too coarse. His areas 17, 18, and especially 19 are all concerned with vision. We now know, as we shall see in the next chapter, that while 17 can be considered a single area, 18 and especially 19 have many significant subdivisions, so that the terminology is no longer used, although it survives with reference to man, in parts of the medical literature.

To summarize, the early parts of the visual system are highly parallel— many similar but distinct neurons are all active at the same time. The retina, at the back of the eye, starts processing the visual input. It sends this information along two major pathways—to the LGN on the

---

*It was Oskar Vogt who cut up and examined Lenin's brain, given to him by the Soviet authorities for this purpose.

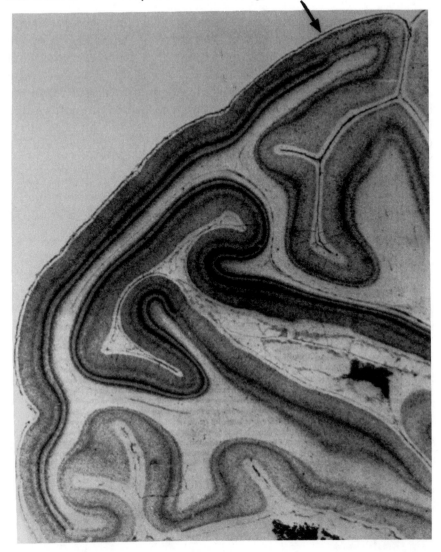

Fig. 44. A section of the visual cortex of the macaque, stained to show the cell bodies. Parts of V1 show the heavy striations (hence the name "striate cortex"). The arrow marks one of the borders between area V1 and V2, which is less striated. The small rectangle has been enlarged in Figure 42.

way to the cortex, and also to the superior colliculus (concerned largely with eye movements)—and to several minor visual regions on the
brain stem concerned with eye movements, the diameter of the pupil,
and the like. Information about color goes to the LGN but not to the
superior colliculus. Any bit of this early information is fairly local and
fairly simple. To allow us to see anything the visual information must
be processed still further in the many distinct cortical areas of the
visual system.

11

The Visual Cortex of Primates

"We should make things as simple as possible, but not simpler."

—Albert Einstein

Each of the two cortical sheets (left and right) can be divided into many fairly distinct cortical areas. How is it decided whether a particular patch of cortex belongs to a single cortical area? There are a number of possible criteria. The first is its structural appearance in cross-section under the microscope—whether it has an extensive layer 4 or not, for example. We have already seen how the striations of area 17 define it uniquely. Such simple distinctions are useful only in a minority of cases, although this may change when more molecular stains become available. Another method of finding the boundaries of a visual area is to examine the details of its visual map, but this method is often inadequate, especially for the higher visual areas, most of which have little or no retinotopic organization—that is, they have no simple visual map. At present, the most powerful method is to find the characteristic patterns of its connections—its inputs and outputs—for each putative area. This can now be done fairly reliably with modern biochemical methods, although, as we saw in Chapter 9, most of these methods cannot be used on human brains.

Many scientists have contributed to this functional division of the cerebral cortex, especially for the cat and the macaque monkey. However, what we know is still incomplete and many of the details must be regarded as tentative.

Let us start with the striate cortex (area 17), now called V1—the first visual area. This is fairly large and is an exception to the rule that there are about 100,000 neurons beneath each square millimeter of cortical surface. In V1, the number is nearer 250,000. In all, the macaque V1, on one side, probably contains about 200 million neurons. This should be compared to the million or so axons coming into it from the LGN. From these numbers we can see immediately that there must be a lot of processing of the inputs from the LGN going on in V1. The high surface density implies that the neurons are, on average, somewhat on the small side, since the thickness of V1 is no greater than that of the adjacent area V2 in which the surface density is lower. One has the impression that evolution has tried to cram as much into V1 as is reasonably possible.

The input from the LGN (which is excitatory) goes mainly into layer 4, although some goes to layer 6. Layer 4 has several subdivisions. The inputs from the P and the M layers of the LGN tend to be segregated into different sublayers of layer 4. All the incoming axons branch extensively, so that one axon may contact as many as a thousand distinct neurons. Conversely, any particular neuron in layer 4 receives inputs from many distinct incoming axons. In spite of this, only a fraction (perhaps 20 percent) of the synapses of a typically spiney stellate neuron in layer 4 receive input directly from the LGN. The rest of the synapses receive input from elsewhere, mainly from the axons of other neurons in the neighborhood. Thus the layer 4 neurons not only listen to what the LGN is telling them but are engaged in extensive conversations with each other about it.

Just as the retinal input is mapped onto the LGN, so the LGN input is mapped onto V1. It is, of course, a map of the opposite half of the visual field, but the map is not uniform (Fig. 45). Much more space is devoted to regions near the center of gaze than to the visual periphery. It reminds me of one of the humorous maps, popular years ago, depicting a New Yorker's view of the United States. Most of it was devoted to Manhattan. New Jersey was scaled down considerably, while California and Hawaii got only a passing indication in the far distance.

In addition, the cortical map, on a small scale, is surprisingly patchy.

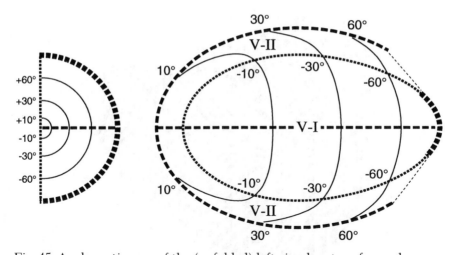

Fig. 45. A schematic map of the (unfolded) left visual cortex of an owl monkey. Only V1 and V2 are shown. The little figure on the left shows the right
visual field. Note the symbols used for its various parts. These symbols are
repeated in the map on the right. The center of the visual field—the first 10
degrees or so—occupies a large cortical area compared to the periphery, from
60 to 90 degrees. Also note how the representation of V2 is split.

Where there is a connection through the LGN from both eyes—that
is, everywhere except from the blind spot and the far periphery—the
two connections into layer 4 are segregated into irregular stripes,* not
unlike fingerprints (Fig. 46). Along the centers of these stripes, but in
the layers above and below layer 4, run a series of "blobs," shown up by
staining for a certain enzyme (cytochrome oxidase). The neurons in
them appear especially interested in color and brightness.

Generally speaking, different neurons in cortical area V1 are interested
in different things. Recall that the input from the LGN comes from
neurons with small receptive fields that are structured as centers with

---

*The exact pattern of stripes and blobs is crudely similar but not identical in
detail from monkey to monkey of the same species. Even for a single monkey, the
pattern differs from one side of the brain to the other, just as the fingerprints on your
left hand are not exactly the same as those on your right hand, and for the same reason: The details depend somewhat on the accidents of the developmental processes.
Once again we are confronted with a situation that has some degree of order but is
distinctly messy in detail.

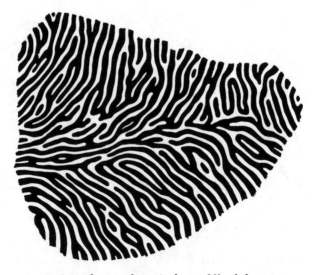

Fig. 46. A reconstruction of part of cortical area V1 of the macaque monkey.[1] The black regions get input from one eye; the white regions from the other eye. This makes the map, shown in Figure 45, somewhat jumbled on a small scale.

antagonist surrounds. Some of the neurons in layer 4 of the macaque still have this property, although their receptive fields are a little larger. As discovered in the sixties by David Hubel and Torsten Wiesel (then both at the Harvard Medical School), most of the other neurons in area V1 respond best to a thin bar of light (or darkness) or to an edge, rather than to a spot of light. For this and other discoveries, Hubel and Wiesel were awarded a Nobel Prize in 1981. The response is better to a moving line than to one flashed on or off. For any particular neuron there is one particular *orientation* of the line or bar to which it fires most vigorously. If the orientation deviates from this by as little as 15 degrees, the firing rate is usually much slower. Different neurons have different preferred orientations, although those immediately above or beneath each other (except those in some parts of layer 4) tend to respond to the same orientation. This arrangement is often referred to as "columnar." Moreover, as one moves in a horizontal direction through the cortex, this preferred orientation tends to change fairly smoothly, although there can be occasional sharp discontinuities. Over any small area of cortex, about 1 millimeter in diameter, the receptive fields of all various neurons tend to overlap somewhat and, at the same time, all the possible orientations are represented. This arrangement has been described as a "hypercolumn" and also as a "cortical module," but the latter idea should not be taken too literally. It is unfortunately much too popular with theorists, including some who should know better.

Hubel and Wiesel discovered that there are two broad classes of selectively oriented cells, which they called "simple" and "complex." Simple cells have well-defined excitatory and inhibitory subregions of their receptive fields, so arranged that the cell responds best to a line or an edge. Some of these fields are of a much finer scale than others and thus respond to finer details.*

Complex cells are different from simple cells in that they do not appear to have neat subdivisions of their receptive fields into separate excitatory and inhibitory regions. To fire well they need a line or an edge that has the preferred orientation and also lies in their receptive field, but they are not fussy as to exactly where the line is located in that field. Their receptive fields tend to be larger than nearby simple cells, but not much larger. In addition, some of them can fire to more complex stimuli, such as a pattern of dots all moving in the same direction.

It is sobering to realize that after almost thirty years of research we still do not know for certain how either simple cells or complex cells are wired to produce their observed behavior. In logical terms the problem seems straightforward. A simple cell will only fire if most of a set of points (constituting the preferred line) *add up* to produce a response. They perform an AND operation, but need a certain threshold input to fire. A complex cell, on the other hand, will fire if either this line or that line or the other line are present (all lying somewhere within its receptive field and having similar orientations). Thus it appears as if a complex cell receives its input from a whole set of similar simple cells and performs an OR operation on them. It is true that the complex cells appear slightly further on in the processing than the simple cells, but on closer examination this straightforward idea leads into difficulties since many of them have some direct inputs from the LGN. In addition there is the problem that the best response is usually

---

*Much fuss has been made about the possibility that such neurons act to produce a Fourier Transform of the visual scene. Taken literally, this is absurd. In any case, they conform better to Gabor Transforms, but it has yet to be established that even this idea will be of any real use. What is certain is that some neurons respond best to fine details (or "spatial frequencies" as they are often called) while others respond better to intermediate, or to coarser, details.

to a *moving* line, and that sometimes a neuron will much prefer movement (perpendicular to the line) in one direction than in the opposite one.

It is particularly unfortunate that this problem has not been solved. There is at least a chance that the AND type of operation performed by simple cells, followed by an OR operation by complex cells is a general strategy used by all areas of the cerebral cortex. If this were the case, it would be most important to know it.

The neurons in cortical area V1 respond in varying ways. As we have seen, many of those in layer 4 are of the center-surround type. This is also true for the neurons in the blobs. Most of the other neurons are orientation-selective, although some respond best if the line is not too long* (referred to as "end-stopping") while others, such as many of those in layer 6, respond best to very long lines.

Another type of neuron gets input from both eyes but will only fire strongly if its input is derived from retinal neurons that are not in exactly corresponding positions in the two retinas. This is needed if the brain is to extract information about the distance of the object in the visual field, since objects at different distances produce different disparities (as already explained in Chapter 4). As we have seen, some neurons are interested in a particular direction of movement, and do not respond to movement in the opposite direction. Many of these neurons lie in a thin layer called 4B. Many neurons respond in much the same way to light of any visible wavelength, while for others, especially those in the blobs, the response of the center and the surround of their receptive field may be selectively sensitive to wavelength. In short, they are interested in color. All this shows that different neurons in V1 are processing the incoming visual information in different ways.

The receptive field is that part of the visual field where a change of light will make the cell fire. There is, however, a much larger area surrounding the receptive field within which changes of light will not by themselves make the cell fire but can modulate the effects produced by the receptive field proper. This large area is now called the "non-classical" receptive field. It introduces the important idea of local context. This context can be feature-specific. The cell is not only interested in a particular feature but is also influenced by similar features in the neighborhood. This very significant aspect of neuronal

---

*They may be involved in producing the kind of illusory contours formed by line terminators, as shown in Figure 15.

behavior probably occurs at all levels in the visual hierarchy. It is likely to have important psychological implications, since psychologists find in many situations that context is important.

Why should cortical area V1 have a map, albeit a crude and distorted one, of the visual field? It is not because there is a little person (the homunculus) looking at it—our Astonishing Hypothesis would forbid this. The most likely reason is that it keeps the brain's wiring shorter. Since a neuron in V1 is mainly concerned with what is happening in only a small region of the visual field, an approximate map keeps each neuron fairly close to those with which it must interact to extract the information it expresses. Theorists have pointed out that this requirement for minimal wiring may also explain the various kinds of patchiness found in the cortex, since it allows multiple submaps to exist within one overall, main map.[2] One little patch of one submap can have strong interactions within it, together with somewhat longer connections to other, nearby parts of the same submap. Such a patch may also have weaker, local connections to the adjacent parts of other types of surrounding submaps. In the same way, a city can sometimes be usefully regarded as being made up of many interacting local communities and associations of common interest, their arrangement being dictated partly by ease of communication. Hence not one but many supermarkets are scattered locally all over the city so that a city dweller is never far from one of them.

This question of economy of wiring will eventually need to be addressed at all levels. Together with the necessity of keeping the total number of neurons in the neocortex to a decent minimum, it may well explain the general nature of cortical organization and of the visual system in particular.

The map in area V1 (and those in other regions) is constructed in such a way that it seems likely that its broad features—which region of V1 corresponds to the fovea, for example—are probably laid down during the development of the brain, largely guided by the actions of the relevant genes. The finer details of the map are produced by modifications made by input from the eyes and seem to depend on whether the firing of the various incoming axons are correlated with each other or not. Some of these developments may occur even before birth. There is a critical period in the life of the very young animal when such alterations to the wiring are probably made fairly easily but some modifications to the map can be made later in life.

*   *   *

It is useful to have a general phrase that expresses the specificity of the response of a neuron, such as the response of many neurons in area V1 to orientation. The one often used is "feature detector." This does indeed capture the fact that some neurons are sensitive to orientation, some to disparity, some to wavelength, and so on. However, the phrase has two shortcomings. The first is that it implies that the neuron responds *only* to the "feature" after which it is named. (Some might assume that it is the only neuron that responds to that feature, which is far from the case.) This overlooks the fact that it may also respond to other (usually related) features. For example, an orientation-sensitive cell with end-stopping responds well to a short line (of the right orientation and in the right place), but because of the substructure of its receptive field, it can also be sensitive to the curvature of a much longer line, part of which lies within its receptive field.

The second misleading aspect of feature detector is that it implies that the neuron is being used by the brain to produce an awareness of that particular feature. This may not be the case. For example, a neuron that responds differently to different wavelengths does not have to be an intimate part of the system that allows you to see color. It might be part of another system that only draws the brain's attention to color differences without producing awareness of what the color actually looks like.

Another aspect of the features coded by feature detectors that is rarely commented upon is that they seldom fall into neat classes as they would if an engineer had designed them. For example, the "simple" type of orientation-selective cells might have been expected to have their excitatory and inhibitory subfields arranged in two ways, one symmetrical about the long axis of the receptive field and the other antisymmetric.* While such types do occur, so do all sorts of other related but messy arrangements. As we shall see in Chapter 13, this is exactly what one expects if they have been developed as part of a neural network, using a built-in learning algorithm (a learning rule) rather than if they had been rigidly laid down in advance by a designer.

To understand what part a particular neuron plays in the operations of the brain, we need to know, as a minimum, not only its receptive field but also what its output projects *to*—that is, all the neurons to which its axon makes synaptic contacts. Terry Sejnowski (now at the Salk Institute) has called this its "projective field" by analogy to the

---

*The first corresponds to a damped cosine wave, the second to a damped sine wave.

term *receptive field*. The projective field will probably play an important role in any discussion of "meaning." It is unlikely that the activity of a neuron whose axon has been severed can have much meaning for the brain.

Cortical area V2, the second visual area, is also a large area and, like V1, it has a "map" of the opposite half of the visual field. If the map of V1 looked a little unusual, because the local scale (called the "magnification factor") varied from the foveal part to the periphery, the map of V2 is even more peculiar, as can be seen from a careful scrutiny of Figure 45. The map is essentially split into two parts, corresponding roughly to the upper and lower parts of the contralateral half of the visual field.* Again there is more area devoted to the regions near the fovea than to the peripheral regions of the visual field.

The V2 neurons, as a whole, are interested in much the same general properties as those in area V1, such as orientation, movement, disparity, and color—but there are differences. Almost all the neurons in V2 receive inputs from both eyes. Their receptive fields are, however, usually bigger than those of V1 and can respond in more subtle ways. For example, neurons have been found that fire to certain subjective contours.† Although neurons that respond to the line-terminator type of subjective contour (Fig. 15) have been found[3] in cortical area V1, neurons sensitive to the other type (the line-continuation type, Fig. 2) have not been found in V1 but do occur[4] in V2. At least one philosopher was astonished to learn that there were neurons that responded to subjective contours but it need not surprise us. It is probably a good general rule that when we see some visual feature *explicitly* (as opposed to merely inferring it), there will be neurons in some areas of our brains that are firing to it. If this rule turns out to be true, it will be an important one.

Cortical area V2 is also patchy, but the enzyme that shows up the blobs in V1 now shows up rather ragged stripes running roughly perpendicular to the V1/V2 border. Careful studies of the neurons in these stripes show that the general visual features they respond to are not the same in each type of stripe. There appear to be several distinct streams of information flowing through V2. One deals mainly with

*It helps to follow the markings that show the positions of the center of gaze and of the horizontal and vertical meridians of the visual field (see the left-hand part of Fig. 45) on the (flattened) surface of the cortex.

†Subjective contours, also called "illusory contours," are faint lines that we see although they are not actually in the visual field, as in Figures 2 and 15.

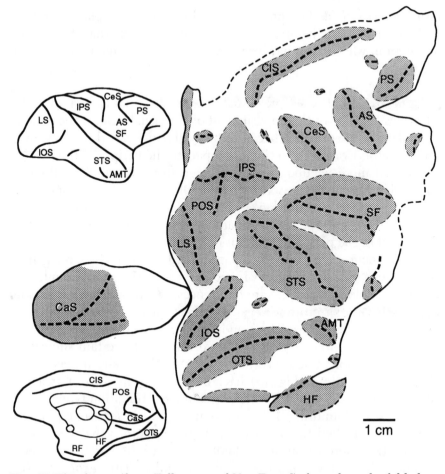

Fig. 47. This figure (from Felleman and Van Essen[5]) shows how the folded macaque cortex is unfolded (usually mathematically, using sections of the brain) so that its arrangement can be grasped more easily. Two views are shown of the unfolded cortex on a smaller scale. The one at the top left shows the view of the right-hand side of the macaque's brain, viewed from the outside. The one on the lower left shows what it would look like from the inside if the brain were cut in half. The lines mark the various infoldings, with initials for their names (for example, *PS* stands for Principal Sulcus).

    The main figure shows the result of unfolding the cortical sheet. The heavily dashed lines represent the depths of each sulcus. The regions folded inside, and thus not on the overall surface of the brain, are shaded. To reduce the distortion produced by unfolding the curved sheet some cuts have been made in the sheet. There is one all around the first visual area, V1 (sticking out on the left), and two others.

color, another mainly with disparity, and so on. All these details are of intense interest to scientists because they bear upon the exact way the various neurons, in the various subregions, should be classified and how they help us to see. The important point for us is that the behavior of the neurons, even in a single area, is segregated into partly disjoint classes, although it can be debated exactly how clean the segregation is.

So far I have talked only about neurons in V1 that project to V2. Are there any neurons in V2 that project back to* V1? The answer is that almost as many neurons project backwards from V2 as project forward from V1, but there is an important difference. The forward projection goes heavily into layer 4 of V2, whereas the backward projection to V1 avoids layer 4 altogether.

Historically, there were considered to be only three visual cortical areas: 17, 18, and 19. I have described two areas in more detail: V1 (equivalent to 17) and V2 (part of the old 18). How many more are there? Surprisingly, at least twenty distinct visual areas have been identified, plus about seven more that are partly visual. This fact alone makes clear the complexities of the visual process. The behavior of the neurons is significantly different in each area, since each area has a distinct set of inputs and outputs. Figure 47 shows how David Van Essen, now at Washington University, has constructed a flattened version of the macaque cortex. Since the cortex is both curved and folded, this necessarily produces some distortion of the map.† The distortion has been reduced by putting a few selected cuts in the cortical sheet, including one that almost isolates area V1, shown sticking out on the left of the figure. This figure should be compared with the next one, Figure 48, in which the marks indicating the cortical folds have been omitted and in their place the many cortical areas have been drawn in. Those that are visual (plus those partly visual) have been shaded. For the macaque, they make up a little more than half the total cortical area (recall that monkeys are very visual animals).

This map is by no means final. For example, area 46 (top right) may yet have to be subdivided. Many areas have been given fancy names, but they are usually referred to by their initials (MT stands for middle

---

*I say "back to" because the broad flow of information, from the retina to the LGN, to V1, and thence to V2, is conventionally regarded as "forward." Workers in Artificial Intelligence usually use the term *bottom-up* instead of forward. The flow of information in the opposite direction they call "top-down."

†In mathematical terms, the Gaussian curvature in some places is far from zero.

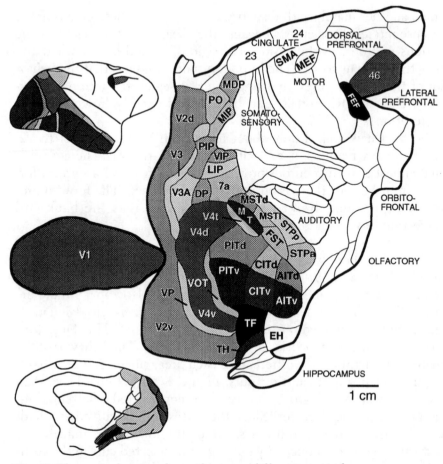

Fig. 48. The main drawing shows the many different cortical areas on one side (here the right-hand side) of the macaque's brain. The two smaller figures (on a smaller scale) on the left show the view from the outside (the upper one) and from the inside (lower one), as if the brain were cleaved in half. The cortical sheet has been unfolded (as described in Fig. 47).

The many areas connected with vision are shaded. Their various names, abbreviated in most cases to initials, are shown in the figure. Their interconnections are shown in Figure 52. The main flows of information are broadly from V1 (on the left) toward the areas on the right of the figure, especially those on the bottom right.

temporal; VIP for ventral intraparietal, etc.). Others have numbers (omitted here) usually corresponding to the numbers given by Brodmann, although some, like 7a and 7b, have been subdivided.

Rather than relate everything that is known about all these visual areas, especially since, for many of them, the information is rather skimpy, let me briefly describe two of them: MT and V4. Cortical area MT (sometimes called "V5") is a small one. It has a fairly good retinotopic map of the visual hemifield, but the receptive fields of the neurons are typically bigger than those in either V1 or V2. Its neurons are strongly interested in the movement of stimuli, including direction of movement. Each neuron will fire to a range of stimulus velocities. Some fire best for fast movements, others for slower movements.

It was not at first appreciated that the response of these neurons often depends on the movement of an object relative to its background. John Allman, of Caltech, arrived at this idea because, unlike many neuroscientists, he is very interested in monkeys and their way of life in the wild. Until recently he kept monkeys at home. He has made several trips abroad to study them in their native habitat, so he knows at first hand what their typical visual environment looks like. He therefore tried to reproduce it, in a greatly simplified form, in his laboratory tests. He and his colleagues used as a stimulus a bar on a TV screen made of random dots.[6] A neuron may fire well to a bar of speckled dots in its receptive field moving (perpendicular to its length) toward, say, the upper right of the visual field. However, if there is movement in the same direction of a background of speckled dots, he found that the firing of the neurons is reduced. If this background moves in the opposite direction, the firing of the neuron to the moving bar may be increased. Thus, what the neuron is mainly interested in is the *relative* movement of its local feature against similar features in the neighboring background. This is the simplest case of the nonclassical receptive field referred to earlier. Although things don't always work this straightforwardly,* it looks as if an ensemble of such neurons can learn to respond not merely to one aspect of an object but also to some aspects of the object's context.

---

*More recently, Richard Born and Roger Tootell, at the Harvard Medical School, have shown[7] that in the owl monkey there are two types of neurons in MT, each of which exists in many small columnar clusters. The first type behaves roughly as described in the text. For the second type, the surround is not antagonistic but *enhances* the main response of the neuron.

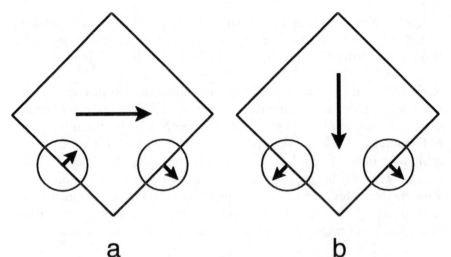

a   b

Fig. 49. *The Aperture Problem.* Consider four lines—the edges of a diamond shape—moving rigidly together, either to the right, as in *a,* or downwards, as in *b*—shown in each case by the large arrow. Each small circle represents a limited aperture through which a neuron is "looking at" the visual field. Through the aperture a single neuron in the early stages of the visual system cannot see which way the diamond is moving. It only senses the movement perpendicular to the small straight line in its visual field, as shown in each circle by the small arrows. The way the diamond is moving can be found by using information from more than a single neuron—compare the direction of the small arrows in *a* and *b.*

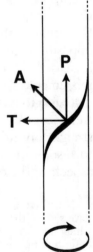

Fig. 50. This illustrates the essence of the barber's pole illusion. The figure shows just one single boundary on the pole.
    The true movement at one point is shown by the arrow marked *T,* since the pole is rotating on its own axis. What would have been seen through a very small circular aperture at that point would be the movement corresponding to the arrow marked A. The brain makes an incorrect synthesis of all the A-type movement information and perceives motion in the direction of the arrow marked *P.* One of the tasks of theorists is to explain just exactly how it makes this mistake.

Some neurons in area MT respond to movement in a more complex way. Their behavior is related to the so-called *aperture problem*. Consider Figure 49. Imagine a small circular hole (the aperture) in a screen through which one views a featureless straight line. This is part of a longer straight line most of which is hidden by the screen containing the aperture. If this rigid line moves in any direction, all you will see through the aperture is a little piece of the line moving *perpendicular to its length*. (This is explained more fully in the legend to Figure 49.)

A neuron in area V1 that is sensitive to direction of movement behaves in just this way. It only senses the component of the movement of a line perpendicular to the line, not the true movement of the whole object. However, some neurons in MT do respond to the true movement, especially when the signal consists of several sets of lines. It would be nice to be able to report that the neurons in MT fell neatly into two classes, those that solve the aperture problem and those, like the neurons in V1, that do not. The truth is far more messy. The neurons show a whole range of behaviors between these two.[8,9] Nevertheless, they are a good example of how the responses of neurons become more sophisticated at higher levels in the visual system.

If the incoming information is misleading, the brain can make the wrong interpretation. A familiar example is the barber's pole illusion—the pole is actually rotating about its long axis but the stripes appear to us to be moving upwards* along its length. Any point on the boundaries between the red and white stripes is actually moving *perpendicular* to the length of the pole. Still, the brain sees the stripes moving along the pole's length. This is illustrated in Figure 50.

The neurons in cortical area MT have little interest in color as such, although some of them can respond to the movement of boundaries produced only by color differences without any difference in luminance across them. This is in marked contrast to the neurons in cortical area V4, which show complicated responses to wavelength but are fairly indifferent to movement.† The receptive fields are typically large, yet in some cases a neuron can respond to a small object, having the appropriate visual properties, everywhere within it. The mapping has some complex retinotopic aspects but does not have a simple retinotopic map in the way V1 does.

---

*Or downwards, depending on the direction of rotation of the pole and the way the stripes are drawn.

†Area V4 is a large one, and, indeed, Van Essen has divided it into three subareas: V4t, V4d, and V4v.

Many of the color responses are "double-opponent responses" that theories of color vision would lead us to expect. More important, the neurophysiologist Semir Zeki, who works at University College, London, has shown[10] that their behavior exhibits the Land effect (already described in Chapter 4). Their response is not just to the wavelengths of the light in the center and the surround of their receptive field, but is strongly influenced by the wavelengths entering the eye from surfaces in the neighborhood. Broadly speaking, they are not just responding to wavelength; they are responding to perceived color. A neuron in the macaque's V4 fired to a red patch, in a pattern of rectangular patches of different colors, whenever Zeki himself saw it as red, even if, by fiddling with the wavelengths of the illumination, the actual wavelengths coming into the retina from that patch had been made very different. This is clearly another case of a neuron's behavior being influenced by relevant context. It is important for psychologists to realize that, to some extent, response to context can be expressed specifically by individual neurons, and that they should allow for this in their theoretical models.

Figure 48 sketches the presently known visual areas but says nothing about how they are connected. Basically, the main flow of information starts from cortical area V1 on the left and proceeds toward those areas farther to the right (i.e., nearer the front of the brain) that border on the nonvisual parts of the cortex. There is often a rough mapping in these projections, meaning that axon terminals that end up near each other in the recipient area usually originate from neurons that are not too far apart in the donor area. This may happen even if there is no retinotopic map in the area, as is the case for the higher areas in the hierarchy.

Van Essen and his colleagues have attempted to arrange all the visual areas in a crude hierarchy using an idea originally suggested by the neuroanatomists Kathleen Rockland and Deepak Pandya. What Rockland and Pandya noted was that if the projection from area A to area B went heavily into layer 4, the reverse projection, from B back to A, largely avoided layer 4 and usually connected strongly to layer 1. We have already seen that this happens for the connections between V1 and V2. This generalization can be symbolized rather simply, as shown by the drawings in Figure 51. The projections in the direction from the eyes to the brain—those going heavily into layer 4—are called the "forward projections" and those in the reverse direction, the "back projections."

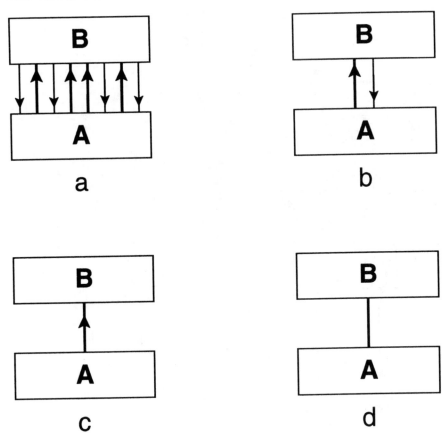

Fig. 51. This illustrates the convention used in Figure 52. It illustrates just two cortical areas, here labeled A and B. There are many connections between them, running in both directions, as shown in *a*. These can be symbolized by just two lines, one running from A to B and the other in the reverse direction, as in *b*. To make it even simpler, the second line can be omitted and just one line shown, as shown in *c*. This is chosen to represent the direction of the primary flow of information. (The flow in the other direction is implied by this single line.) To make it even simpler, the arrow in *c* can be omitted, as shown in *d*. This implies that the main flow of information (the so-called forward direction) is always made upwards in the diagram, so that B must be drawn above A, and not vice versa.

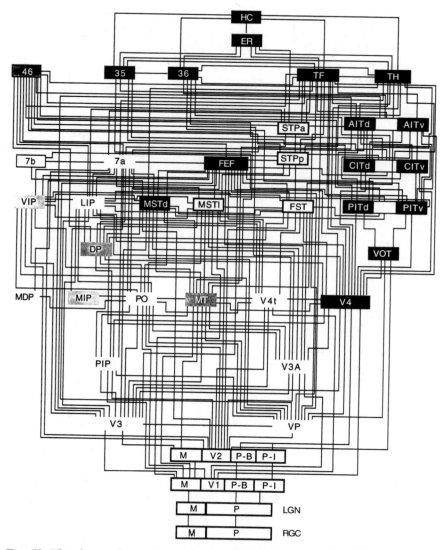

Fig. 52. This figure shows the many connections between the different visual areas. It uses the convention, explained in Figure 51, that each line represents many axons, running in both directions. At the bottom of the figure, labelled *RGC*, are the retinal ganglion cells in the eye. *LGN* is part of the thalamus. This projects to the first visual area, V1, shown subdivided into four parts, with the second visual area, V2 (also in four parts), just above it. The names of the different areas are fairly arbitrary and need not concern the reader. At the top, *HC* stands for hippocampus and *ER* for the entorhinal cortex that leads into it. The arrangement is semihierarchical, as explained in the text. Many other (nonvisual) cortical areas, shown in Figure 48, are not shown in this figure. Modified by Suzuki and Amaral from Felleman and Van Essen.

Is the rule about layer 4 connections always obeyed? There are indeed complications, but it has proved possible to show that most of the known connections, using the convention of Figure 51, can be represented on a single hierarchical map. The latest version is shown on Figure 52. (Recall that each line in this diagram symbolizes very many axons running in both directions.) You should not immediately take fright at all the complicated details of this outline wiring diagram but merely note that it demonstrates, if nothing else does, the complexity of the visual process. Very few people would have guessed that their brains were wired up in this way.

The agreement with the layer 4 convention has some important exceptions. There are, for example, many cross-connections between cortical areas at the *same* level. The simple layer 4 rule does not cover these, so somewhat more elaborate rules are actually used to construct the diagram. It is not yet known whether the true arrangement is really only quasi-hierarchical, or whether the exceptions to these more complex rules are largely due to experimental error. Nevertheless there is no doubt that the various areas can be crudely arranged in an approximate hierarchy. Do the exceptions, if they exist, have a special significance? Only further work can answer that question.

Notice that while most connections to another area are either to the same level or to one level higher or lower, there are well-established exceptions—that is, a connection can skip levels. An example would be the direct connection from area V1 to area MT, which is four levels above it. The rule that all connections are reciprocal is almost always true, but again there are exceptions.* Nor, incidentally, does Figure 52 attempt to show the strength of the connections (for example, how many axons each line represents), mainly because information on this point is at present too sparse. Some lines in Figure 52 represent millions of axons; others as few as 100,000 or less.

Do areas adjacent *in the cortex* always connect to each other? As might be expected, they usually do, but again there are a few exceptions.

The hierarchical arrangement is supported by a somewhat different source of evidence. This is the general nature of the responses of the neurons in the different areas. As we ascend the hierarchy this behavior follows two rough rules: The sizes of the receptive fields increase,

---

*V4 projects back strongly to V1, but the forward projection from V1 to V4 is usually weak or absent.

so that for the highest areas they often cover the whole visual hemi-field and even part or all of the other half of the visual field (connect-ed via the corpus callosum). In addition, the features to which the neurons respond become more complex. Some neurons in area V2 respond to certain subjective contours, while some in area MT respond in a less simple way to patterns of movement (as we have seen, they solve, or partially solve, the aperture problem). Neurons in area MST fire in response to movements over the visual field corre-sponding to light from an object that is zooming nearer, others to light from a receding one. Neurons in V4 respond to perceived color, rather than just to the wavelength of the light.

When we come to the higher areas we find neurons responding to the front view of a face, although they are not too fussy about its posi-tion in relation to the center of gaze, nor even if it is tilted a little. Such a neuron will respond hardly at all to a picture made up of a jum-bled arrangement of eyes, nose, mouth, and so on. Other neurons respond best to a face seen in profile. Neurons in area 7a, on the other hand, are mainly interested in *where* an object is in relation to the head or the body, and much less in *what* it is. The latter is the main concern of the inferotemporal regions (those with IT in the middle of their initials such as CITd) as already described for faces. It is almost certain that many more complex responses remain to be discovered.

The general pattern, then, is that each area receives several inputs from lower areas. (These lower areas have already extracted more com-plex features than the rather simple ones to which area V1 responds.) It then operates on this combination of inputs to produce even more complex features, which it then passes on to the even higher levels in the hierarchy. At the same time, the information flows in somewhat separate but interacting streams up the hierarchy. We have already seen examples of this in the partial segregation of the M and the P sig-nals from the retina, the three streams from V1 to V2, and, at a higher level, the "what" and the "where" flows, but it must be emphasized that there is always some cross-talk between such streams.

And what of the back-pathways? These urgently need more detailed study. One can imagine various functions for them. They might help to produce the nonclassical receptive fields referred to earlier and so allow activity at a higher level to influence responses at a lower level. They might be part of systems that signal to a lower area that its oper-ations have just been successful on a slightly more global level and therefore should be remembered—that is, the synapses should be modified to allow that feature to be more easily detected in the future. They may be closely involved in the mechanism of attention and the mechanisms needed for visual imagery. They may help to synchronize neuronal oscillations (see Chapter 17). These are some of the more obvious possibilities, but which of them, if any, is true remains to be seen.

Moreover the whole system does not look like a one-shot, static, response mechanism. It is likely to operate by many transient, dynamic interactions, conducted at a fairly fast rate. And finally, let us not forget that everything I have described applies to the macaque monkey, and not to human beings. It is reasonable for us to assume that our own visual system is similar to that of the macaque but this is only an assumption. For all we know now, it could differ not only in detail but also in complexity.

The secret of the neocortex, if it has one, is probably *its ability to evolve additional layers to its hierarchies of processing*, especially at the upper levels of those hierarchies. Such extra layers of processing are probably what distinguishes higher mammals, like man, from lower mammals, such as the hedgehog. I suspect the neocortex uses special learning algorithms that permit each cortical area to extract new categories from experience, even though each area is embodied in a complex processing hierarchy. This ability probably distinguishes the cerebral neocortex from other forms of neural architecture, such as the cerebellum and the corpus striatum, that are not hierarchical in this complex way.

These ideas are speculations, yet one thing is fairly clear: Although there are many different visual regions, each of which analyzes visual input in different and complex ways, so far we can locate no single region in which the neural activity corresponds exactly to the vivid picture of the world we see in front of our eyes. It might be thought, by looking at Figure 52, that this occurs in some more highly complex structure like the hippocampus and its associated cortical structures (marked HC and ER), which are located at the top of the hierarchy. But as we shall learn in Chapter 12 a person can lose all these regions (on both sides of his brain) and still report that he sees fairly well and indeed behaves as if he does. In short, *we can see how the brain takes the picture apart, but we do not yet understand how it puts it together.* How does it construct the well-organized and detailed visual awareness of all the objects, and the behavior of these objects, in our visual field?

12

Brain Damage

"Babylon in all its desolation is a sight not so awful as that of the human mind in ruins."

—Scrope Davies

Over the years, neurologists have examined people whose brains have been damaged in various ways—by strokes, blows to the head, gunshot wounds, infections. Quite a number of these injuries alter some aspects of visual awareness while leaving other abilities, like speech or motor activities, more or less intact. The evidence suggests that there is a remarkable degree of functional specialization in the cortex, often in rather surprising ways.

In many cases, the damage to the brain is not very "clean" or specific. A speeding bullet is no respecter of cortical areas. (The living cortex has the texture of a rather soft jelly. Bits of it can easily be removed by sucking on a pipette.) Typically, the damage is likely to involve several cortical areas. The most dramatic effects are produced by damage to corresponding places on the two sides of the head, although such cases are rather rare.

Many neurologists have time only for a short examination of an injured patient—just long enough to be able to make an educated guess as to where the damage is likely to be. Lately, even this form of detective work has been largely superseded by brain scans. In the recent past it was the usual practice to report on a dozen or so similar cases together, since it was felt that to describe a single, isolated injury was unscientific. This unfortunately tended to lump together what were really somewhat distinct types of damage.

Recent trends have corrected these practices to some extent. Particular attention is now often given to the small number of cases in which one particular aspect of perception or behavior is altered while most other aspects are spared. These patients are likely to have suffered more limited and therefore more specific damage. An effort is also made to localize the damage* via brain scans. The living patient, if he is cooperative, can be given a whole battery of psychological and other tests in order to discover just what he can or cannot see or do. In some cases such tests have extended over a number of years. As ideas about visual processing have become more sophisticated, experiments to test these ideas have become more subtle and extensive. They can now be combined with brain scans that record the *activity* in the brain during these different tasks. These results on several patients can be used to compare and contrast patients with either similar damage, or similar symptoms, or both.

The obvious example to begin with would be damage to V1, the striate cortex. If V1 is totally destroyed on one side of the brain, the patient appears to be blind in the opposite hemifield. Toward the end of this chapter a strange phenomenon known as "blindsight" will be discussed in detail. At this point let us look instead at the effects of damage to one of the highest parts of the visual hierarchy, and to the right-hand side of the head only. This is known as unilateral neglect or hemineglect. The damaged area corresponds broadly to area 7a in the macaque's brain (see Fig. 48). It is usually caused by vascular disorders in one of the cerebral arteries, commonly known as strokes.

In its early stages, the symptoms can be extreme—the patient's eyes and head may be rotated to the right. In really severe cases, where the damage may be so extensive that control and feeling have been lost on the patient's left side, he may deny that his own left leg belongs to him. One man was so incensed at having somebody else's leg in bed with him that he threw it out of the bed and was surprised to find himself lying on the floor.

Most cases are not as severe as this. After some days the extreme

---

*The damage, if it is not progressive (as it is in cases of cancer, for example, or of Alzheimer's disease), can in principle also be localized by a detailed examination of the patient's brain immediately after death, but this is often not possible.

symptoms usually lessen or disappear. The patient may, for example, fail to pick up food from the left side of his plate. If asked to draw a clock, or a face, he will typically only draw the right side of it. After some weeks, as the brain partially recovers, the severity of this hemineglect diminishes further, but the patient still appears to attend less to the left side than to the right. If asked to bisect a line he will draw the midway point over to the right. He is not, however, truly blind on the left. He can see an object there if it is in isolation, but he may fail to notice it if there is also something significant on his right-hand side. Moreover, he often denies that anything is amiss and does not report seeing an empty space on the left side of his visual field.

Hemineglect is not limited to visual perception. It can also apply to visual imagination. The classical example was reported by the Italian Edoardo Bisiach and his colleagues.[1] Their patients were asked to imagine that they were standing at one end of the main square of Milan, facing the cathedral, and to describe what they could remember. They reported details mainly of buildings that, from this point of view, were on their right-hand side. They were then asked to do the same, but this time as if they were standing at the opposite end of the square, with the cathedral behind them. They then reported mainly the details on the side that they had previously neglected to mention, again visualized on their right.

Another remarkable form of brain damage produces a partial or complete loss of color vision. The patient sees everything in shades of gray. This is known as "achromatopsia"—a case was reported by Robert Boyle (the "father of chemistry") as early as 1688. In 1987, Oliver Sacks and Robert Wasserman described such a case in *The New York Review of Books*.[2] The patient was a New York abstract painter, Jonathan I., who had been profoundly interested in color, so much so that listening to music had produced "a rich tumult of inner colors." This synesthesia, as it is called, disappeared after his accident, so that music lost much of its appeal to him.

The damage was a result of a rather minor car accident. Jonathan I. probably suffered from concussion but otherwise appeared to be unhurt. He was able to give a clear account of the accident to the police but later developed a bad headache and also amnesia for the accident. The next morning, after stuporous sleep, he found he could not read, although this inability disappeared five days later. He had difficulty distinguishing colors but no subjective sense that colors had altered.

This developed on the next day. While he knew it was a bright sunny morning, the world looked to him, as he drove to his studio, as if it were in some sort of fog. The full reality of his deficit only struck him when he arrived there and saw his own brilliantly colored paintings. They were now "utterly gray and void of color."

Sacks and Wasserman's account of the psychological effects of this cruel deficit is vivid and detailed. Although his problem could be judged no worse than looking at an old black-and-white movie, this was not how Mr. I. felt. Most foods appeared disgusting to him— tomatoes appeared black, for example. His wife's skin seemed to him to be rat-colored, and he could not bear to make love to her. It did not help him if he shut his eyes. His highly developed visual imagination had also become blind to color. Even his dreams had lost their previously vivid coloration.

Mr. I.'s scale of grays was compressed, especially in strong sunlight, so that he could not see delicate tonal graduations. His overall response to the wavelengths of light was normal, except that he had an additional peak of sensitivity in the short wavelength ("blue") region of the spectrum. This may explain why he could not see white clouds against a background of blue sky. He also had trouble identifying faces unless they were close. His vision seemed sharper to him because objects stood out with considerable contrast and clarity, almost like silhouettes. He was abnormally sensitive to movement, and reported, "I can see a worm wriggling a block away." At night he claimed he could see so well that he could read license plates from four blocks away. Because of this he became, in his own words, "a night-person." While wandering about at night his vision was no worse than other people's.

Mr. I.'s loss of awareness of color has interfered rather little with other aspects of his vision, except that the loss has altered his sensitivity to shades of gray and may have produced extra sensitivity to movement. The damage was clearly bilateral, since both halves of the visual field were affected (some cases of achromatopsia affect only one half) and delayed, since the full loss of color awareness took about two days to develop. Were it not for the increased response to the shorter wavelengths of light (blue) it might appear to be a defect in the P system (the one more interested in shape and color) that had left most of the work of seeing to an undamaged M system (the one more interested in movement—see Chapter 10).

Mr. I.'s brain was given both an MRI scan and a CAT scan (although the latter was on rather a coarse scale), but no damage could be seen, so it is still not yet clear that the damage was truly cortical. However, previous cases have shown that achromatopsia usually involves cortical damage at a fairly high level of the human visual system (the ventromedial sector of the occipital lobe).

Another very striking deficit produced by brain damage is the inability to recognize faces, known as "prosopagnosia." One of England's prime ministers in the last century had this difficulty. He even failed to recognize the face of his eldest son. There are many varieties of prosopagnosia, probably because the exact nature of the brain damage varies from patient to patient. The problem is usually not that of recognizing that a face is a face but of recognizing whose face it is, whether that of a wife, a child, or close friends. The affected person often cannot recognize his own face in a photograph, or even in a mirror, although he knows that it must be his face because it winks when he winks. He can usually recognize his wife from her voice, or the way she walks, but not just from seeing her face.

Unless the damage is extreme, he can describe the features of a face—the eyes, nose, mouth and so on—and how they are related to each other. Moreover, his visual scanning mechanism is normal. In some cases a patient can discriminate between faces when asked to match differently lit photographs of unfamiliar faces, yet he cannot say whose faces they are photographs of, even if they were previously well known to him.

Prosopagnosia often accompanies achromatopsia when the latter is bilateral, but it should be remembered that there is no reason why the damage (often due to a stroke) should affect only a single cortical area. Indeed, prosopagnosia can occur in association with other specific defects.

The neurologist Antonio Damasio has made several important contributions to the study of prosopagnosia.[3] The condition is not necessarily limited to difficulties in face recognition. A case was already known of a farmer who could no longer recognize his cows, each of which he previously knew by name. But Damasio went further than this. He and his colleagues showed that in many cases patients could not recognize the individual members of a class of rather similar objects. For example, one patient could easily recognize a car as a car, yet could no longer tell whether it was a Ford or a Rolls-Royce, although he could identify an ambulance or a fire engine, presumably because they were sufficiently distinct from a typical car. A shirt could be recognized as a shirt but not as a "dress" shirt.

Damasio and his colleagues have also discovered that there are some patients who, while they cannot recognize the identity of a face, can recognize the meaning of facial expressions and can also estimate age and gender.[4] Other prosopagnosia patients lack these abilities. These results suggest that different aspects of facial recognition are handled in different parts of the brain.

There is some controversy as to exactly how prosopagnosia and the underlying mechanisms should be described. Damasio stresses that it is not a general disorder of memory, since such a memory can be triggered through other sensory channels (auditory, for instance). Exactly what the mechanisms are in each type of case remains to be discovered.

A remarkable case of a patient lacking awareness of most types of movement has been reported by the psychologist Joseph Zihl and his colleagues.[5] The damage was bilateral, located in several regions on each side of the cortex. When first examined, the patient was in a very frightened condition. This is not surprising since objects or persons she saw in one place suddenly appeared in another without her being aware they were moving. This was particularly distressing if she wanted to cross a road, since a car that at first seemed far away would suddenly be very close. When she tried to pour tea into a cup she saw only a glistening, frozen arc of liquid. Since she did not notice the tea rising in the cup, it often overflowed. She experienced the world rather as some of us might see the dance floor in the strobe lighting of a discotheque.

We all have this trouble on a vastly slower time scale. The hour hand of a clock does not appear to move, yet if we glance at it some time later it is in another place. We are thus familiar with the idea that things can be moving even if we are not directly aware of the movement, but we normally do not have this difficulty within the usual time scales of everyday life. Clearly we must have a special system to detect movement, in and of itself, without having to infer it logically from two distinct observations separated by time.

Detailed tests showed that the patient could detect certain forms of motion, probably corresponding to a severely impaired short-range mechanism, but the mechanism that made more global associations of movement had been disrupted. Her vision had some other defects, mostly connected with motion, but she could see color and recognize faces, for example, and showed no signs of the type of neglect described earlier in this chapter.

There are many other kinds of visual defects produced by brain damage. Two cases have been reported in which the sufferer had lost perception of depth and saw the world and the people and other objects in it as perfectly flat, so that "the most corpulent individual might be a moving cardboard figure, for his body is represented by an outline only." Other patients can recognize an object if seen from a normal straight-on point of view but not from an unusual one, such as looking at a saucepan from directly above.[6]

Two British psychologists, Glyn Humphreys and Jane Riddoch, studied a patient over a period of five years who had a number of visual defects—for example, he had lost his color vision and also could not recognize faces.[7] They showed that his major visual problem was that while he could see the *local* features of an object, he could not bind them together. Therefore, he could not recognize what the object was although he could copy a drawing fairly well, was articulate, and could produce fluent verbal descriptions of things he had known about before his stroke. Such cases are important because they show that a person who has lost some of his high-level vision can still be visually aware at a lower level. This supports the claim that there is no single cortical area that registers everything we can see.

There is one kind of visual defect that is so surprising that people have been known to doubt that it could possibly exist. This is known as Anton's syndrome or "blindness denial." The patient is clearly unable to see but is unaware of that fact.[8] Asked to describe the doctor's tie the patient may say that it is a blue tie with red spots when in fact the doctor is not wearing a tie at all. When pressed the patient may volunteer the information that the light in the room seems a little dim.

At first, such a condition seems impossible. Alternatively the medical diagnosis might be hysteria, which is not very helpful. But consider the following possibility. I have often found that when talking to someone over the telephone whom I have not met I spontaneously form a crude visual image of his or her appearance. I held several long telephone conversations with one man whom I must have pictured in his fifties, rather thin, with rimless glasses. When, eventually, he came to see me I found he was in his thirties and was decidedly fat. It was only my surprise at his appearance that made me realize that I had previously imagined him otherwise.

I suspect that a person suffering from blindness denial produces such images, probably because the brain damage is such that these images do not have to compete with the normal visual input from the eyes. In addition, due to damage elsewhere, they may have lost the critical faculty that would normally alert them that something was wrong. Whether this explanation is correct remains to be seen, but at least it makes the condition appear not totally incomprehensible.

Are there any trends in the way the different cortical areas react to damage? Damasio has pointed out that in the human temporal region (at the side of the head) brain damage toward the back of the head produces effects of a different character from those produced by damage that occurs farther forward.[9] Damage located toward the rear of the temporal region (or in the occipital region just behind it—see Fig. 27) concerns rather general ("categorical") matters. As one proceeds forward the effects of damage become rather less general till, near the hippocampus, the loss concerns very individual ("episodic") events. Thus the distinction* between categorical and episodic memory may be too sharp a distinction. There may be a gradual transition from areas that deal with general objects and events to those that deal with unique ones.

This suggestion of Damasio's is very much in line with my description (on page 158) of how a cortical area functions. Each area constructs new features from the combination of features already extracted by other areas (usually lower in the hierarchy) that feed into its middle layers.

As one ascends the visual hierarchy, for example, one goes from cortical area V1, which deals with rather simple visual features (such as oriented lines) that occur all the time, to areas that deal with objects like faces, which occur much less frequently, until one reaches the cortex associated with the hippocampus (at the top of Fig. 52), where the combination of signals it responds to (both visual and others) corresponds largely to unique events.

What we've covered is enough to establish two general points: The visual system works in strange and mysterious ways, and its behavior is not incompatible with what scientists have discovered about the wiring and behavior of the macaque's visual system and, by extension, our own.

But our task is to understand visual awareness. Awareness is the result of the many intricate processes needed to construct the visual image. Are there forms of brain damage that bear more directly on awareness itself? It turns out there are several.

*This is mentioned earlier in Chapter 5 and later in this chapter.

The first is often referred to as "split brains." In its cleanest form it results when the corpus callosum—the large tract of nerve fibers that connect the cerebral cortex on one side of the head to that on the other side—is severed completely, together with a smaller fiber tract called the "anterior commissure." This surgical operation is performed to relieve certain cases of epilepsy that have failed to respond to other treatments. The corpus callosum can be lost due to other forms of brain damage, but there is usually some additional destruction elsewhere in the brain, so the interpretation of the results may not be so straightforward. There are also people who are born without the corpus callosum, but the brain usually develops in such a way as to compensate to some extent for early deficits, so again the results are not so dramatic as in the surgical cases.

The history of the subject is so odd that it is worth a brief mention.[10] A distinguished American neurosurgeon reported in 1936 that no symptoms followed if the corpus callosum was cut. Another expert in the mid-fifties, reviewing the experimental results, wrote that "the corpus callosum is hardly connected with psychological functions at all." Karl Lashley (a clever and influential American neuroscientist who, curiously enough, was almost always wrong) went so far as to suggest, partly in jest, that the only thing the corpus callosum did was to keep the two hemispheres from collapsing into each other. (The corpus callosum appears to be somewhat hard, hence the name "callosum.") These opinions we now know to be gross errors, caused partly because the callosum was not always severed completely but mainly because the tests used were either insensitive or inappropriate.

The situation was dramatically changed by the work of Roger Sperry and his colleagues in the fifties and sixties. For this work Sperry was awarded a Nobel Prize in 1981. They showed clearly, by carefully designed experiments, that a cat or a monkey whose brain had been split could be taught in such a way that one hemisphere learned one response while the other hemisphere learned another, or even a conflicting response to the same situation.[11] As Sperry put it, "It is as if the animal had two separate brains."*

Why is this? For most right-handed people only the left hemisphere can speak or communicate through writing. It also rules most of the capacity to deal with language, although the right hemisphere may

---

*These results on animals led to a more careful examination of split brain patients, especially by Sperry, Joseph Bogen, Michael Gazzaniga, Eran and Dahlia Zaidel, and their colleagues.

understand spoken words to a limited extent and probably deals with the music of speech. When the callosum is cut, the left hemisphere sees only the right half of the visual field; the right hemisphere, the left half. Each hand is mainly controlled by the opposite hemisphere, although the other hemisphere can produce some of the coarser movements of the hand and arm. Except under special conditions, both hemispheres can hear what is being said.

Immediately after the operation the patient may experience various transient effects. For example, his hands may act at cross purposes, one buttoning up his shirt while the other follows unbuttoning. Such behavior usually subsides, and the patient appears comparatively normal. But careful testing reveals more.

In the test, the patient is made to fix his gaze upon a screen onto which an image is flashed to one or the other side of his fixation point. This ensures that the visual information will reach only one of the two hemispheres (more elaborate methods are now available to do this).

When a picture is flashed into the patient's left (speaking) hemisphere, he can describe it the way a normal person can. This ability is not limited to speech. When asked, the patient can also point to objects with his right hand (largely controlled by the left hemisphere) without speaking. His right hand can also identify objects by touch even though he is prevented from seeing them.

If, however, a picture is flashed into the *right* (nonspeaking) hemisphere, the results are quite different. The left hand (largely controlled by this nonspeaking hemisphere) can point to and identify unseen objects by touch, as the right hand could do previously. But when the patient is asked to explain why his left hand behaved in that particular way, he will invent explanations based on what his left (speaking) hemisphere saw, not on what his right hemisphere knew. The experimenter can see that these explanations are false, since he knows what was really flashed into the nonspeaking hemisphere to produce the behavior. This is a good example of what is called "confabulation."

In short, one half of the brain appears to be almost totally ignorant of what the other half saw. A little information can sometimes leak across to the other side. While flashing a series of pictures into the right hemisphere of a woman, Michael Gazzaniga slipped in one of a nude. This made the patient blush. The left hemisphere was quite unaware of what the picture was but knew that it had produced a blush, so it said, "You do show some funny pictures, don't you, Doctor?" After a while such a person may learn to make one side cross-cue the other; for example, by signalling in some way with the left hand so that the speaking hemisphere can pick up the signal. In a normal subject, the detailed visual awareness in the right hemisphere can easily be transferred to the left hemisphere so that the person can describe it in words. When the corpus callosum is fully cut, this information cannot cross to the speaking hemisphere. *The information is unable to pass via the various connections lower down in the brain.*

Notice that I am not concerned here with how the two halves of the brain differ, except that language is normally on the left. I do not need to worry about whether the right-hand side has somewhat special properties, such as being rather better at recognizing faces, for example. Nor will I consider the extreme view expressed by some that while the left side is a "person," the right-hand side is merely an automaton. Obviously the right side lacks a well-developed language system and is therefore in some sense less "human" since language is a unique ability of human beings. Eventually we shall need to answer the question whether the right-hand side is more than an automaton, but I feel that can wait till we understand better the neural basis of awareness, to say nothing of the question of Free Will. The balance of professional opinion is strongly of the view that, apart from language, the cognition and motor capacities of the two sides, while not exactly the same, have the same general character.

Most split brain operations do not sever the intertectal commissure (mentioned in Chapter 10) that connects the superior colliculus on one side to that on the other. The brain cannot use this intact pathway to transmit the information in visual awareness from one side to the other. For this reason it is unlikely that the superior colliculus is the seat of visual awareness, even though it is involved in visual attention.

Another fascinating phenomenon is known as "blindsight," studied extensively by the Oxford psychologist Larry Weiskrantz.[12] Patients with blindsight can point to and differentiate between certain simple objects, while at the same time denying that they see them.*

Blindsight is usually caused by fairly extensive damage to the primary visual cortex V1 (the striate cortex), in many cases only on one side of the head. In the test, a horizontal row of small lights is arranged so that when the patient fixes his gaze on a spot to one side of the lights, they all fall into the blind part of his visual field. After the sound of a warning buzzer one of the lights is lit for a short time. The patient is asked to point to where the light was, without moving

---

*There is much parallel work on monkeys that I shall not attempt to describe.

his eyes or his head while the light is on. The patient normally demurs, saying that since he is blind there, there is no point in doing the test. With a little persuasion he may be coaxed to give it a try and "guess" where it was. The test is then repeated many times, sometimes with one light going on, sometimes another. And the surprised patient, in spite of denying that he can see anything, points to the active light fairly accurately, usually within 5 or 10 degrees.*

Some patients can distinguish simple shapes, such as an X from an O, provided they are big enough, and some can also discriminate line orientation and flicker. There are claims that two patients adjusted their grasp so that it matched the shape and size of the object they were reaching for, at the same time denying they saw it. In some cases a patient's eyes can follow the movement of moving stripes, but this task may be handled by another part of his brain, such as the superior colliculus. A patient's pupils can respond to light changes, since changes of pupil size are involuntary and controlled by another small brain region.

Thus, even with a badly damaged V1, the brain can detect some fairly simple visual stimuli and act on them although the affected patient will firmly deny his awareness of them.

What neural pathways are involved is still unclear. It was originally surmised that the information went through the superior colliculus, a part of the "old brain." It now seems unlikely that this is the whole story, since recent experiments have shown that the blindsight responses to wavelength involve the cones of the eye. The response to different wavelengths is similar to that of normal people, although a brighter light is needed.[13] This makes it unlikely that the colliculus is the only pathway, since no color-sensitive neurons have been found there.

The problem is complicated in that the damage to cortical area V1 produces, in time, extensive cell death in the corresponding parts of the LGN (the thalamic relay), and that this in turn kills many of the

---

*Naturally this result was greeted with some skepticism. One objection, for example, was that the behavior was due to light scattered by the eye into places in the retina corresponding to the parts of the visual field that the patient could see. This is unlikely, especially as it has now been shown that a light shone onto the blind spot does not produce the effect. (Recall that there are no photoreceptors in the blind spot, so there is nothing there that can respond to the light. In a blindsight patient, on the other hand, the photoreceptors are intact and can pick up signals. It is the visual cortex that has the primary damage.) All of these objections have now been answered by further experiments, and there is now little doubt that blindsight is a real phenomenon.

retinal ganglion cells of the P type since, like hermits, they have no one to talk to.* However, some P neurons remain, as do some neurons in the relevant regions of the LGN, presumably because they project to some undamaged places. There are direct though weak pathways from the LGN directly to areas in the cortex beyond V1, such as V4. These pathways may remain sufficiently intact to lead to a motor output (being able to point, for instance), but not enough for visual awareness. (See the discussion of Libet's work in Chapter 15.) There is suggestive evidence that in some cases there are little islands of intact tissue within the damaged patch of area V1, so that V1 can still produce some effects in those regions, even though they may be small.[14] Or it may simply turn out that an intact V1 is essential for awareness for another reason, and not merely because it normally produces a strong input to the higher visual areas. Whatever the reason, the patient can use some visual information while denying that he sees anything.

Another interesting form of behavior has been found in victims of prosopagnosia. While hooked up to a lie detector and shown sets of both familiar and unfamiliar faces, the patients are unable to say which faces were familiar, yet the lie detector clearly showed that the brain was making such a distinction even though the patients were unaware of it.[15] Here again we have a case where the brain can respond to a visual feature without awareness.

The hippocampus is a part of the brain that is not confined to vision but is concerned with a type of memory. It is marked HC at the top of Figure 52, which also shows its connection to a part of the cortex called the "entorhinal cortex" (ER in the figure). This has fewer layers than most neocortex. Because of its position near the top of the sensory processing hierarchy, one would be tempted to guess that here at last was the true seat of visual (and other) awareness. It receives input from many of the higher cortical areas and projects back to them. This elaborate one-way pathway is reentrant—that is, it returns very close to where it started—and this also might suggest that this is where consciousness really resides, since the brain might use this pathway to reflect on itself.

The experimental evidence argues strongly against this otherwise attractive hypothesis. Hippocampal damage can be caused by an infection of the virus herpes encephalitis, which produces severe but

---

*Neurons often die if all their outputs lie only on dead neurons.

sometimes rather limited damage. The virus appears to prefer to attack the hippocampus and its associated cortex. The borders of the damage can be quite distinct. Because the damage can be located by an MRI scan and is not progressive, a patient can be followed over a period of years after the acute phase of the infection is over.

If you were to meet a person who had lost his hippocampus on both sides, plus the immediately adjacent cortical areas, you might not at first realize that he was in any way abnormal. It is striking to see a videotape of such a person: talking, smiling, drinking coffee, playing checkers, and so on. Almost his only problem is that he cannot remember any recent episode that happened more than a minute or so before. Introduced to you he will shake your hand, repeat your name, and make conversation. But if you go out of the room for a few minutes and then return, he will deny he had ever seen you before. His motor skills are preserved and he can learn new ones, which usually last for at least several years if not indefinitely, but he cannot remember the occasions when he learned them. His memory for categories is intact, but his memory for new events lasts only a very short time and is then almost totally lost. He may also be handicapped in his memory for episodes that took place before his brain damage. In short, he knows the meaning of the word *breakfast* and he knows how to eat his breakfast, but he has not the faintest idea of what he had for breakfast. If you ask him he will either tell you he can't remember or he will confabulate and describe what he thinks he might have eaten.

Although in a sense he has lost full human "consciousness," his short-term visual awareness appears to be unaltered. If it is impaired, it can only be in subtle ways that testing has not yet uncovered. The hippocampus and its closely associated cortical areas are thus not necessary for visual awareness. It is possible, however, that the information flowing in and then out of the hippocampus may normally reach consciousness, so it is sensible to keep an eye on the neural areas and pathways involved, as this may help to pin down the location of awareness in the brain.*

The study of brain damage sometimes gives us results we can obtain in no other way. Unfortunately, this knowledge is often tantalizingly ambiguous because in most cases the damage is so messy. In spite of

---

*The exact function of the hippocampal system, and the precise way its neurons carry it out, are a matter of current controversy, but the broad picture I have painted would be widely accepted, although there would probably be the usual quibbles about the terminology.

this limitation, in favorable cases the information can be decisive. At the least, the results of brain damage can suggest ideas about the workings of the brain that can be explored by other methods, either on man or on animals. In some cases it confirms for man what we have already learned from experiments on monkeys.

13

Neural Networks

"... I believe the best test of a model is how well can the modeller answer the questions 'What do you know now that you did not know before?' and 'How can you find out if it is true?'"

—James M. Bower

Neural networks are assemblies of variously interconnected units. Each unit has the properties of a much simplified neuron. Neural networks are used to simulate what goes on in parts of the nervous system, to produce useful commercial devices, and to test general theories of how brains work.

Why do neuroscientists need theory at all? If they understand how single neurons behave surely it should be possible to predict the performance of interacting groups of neurons. Unfortunately, this is not as easy as it might seem. Apart from the fact that the behavior of a single neuron is often far from simple, neurons are almost always connected together in intricate ways. In addition, the whole system is usually highly nonlinear. Put in its simplest form, a linear system is one in which twice the input gives exactly twice the output—that is, the output is proportional to the input.* For example, when two sets of small, travel-

*More precisely, y is related to x in a linear fashion if $y=ax+b$ where a and b are constants.

ling ripples meet each other on the surface of a pond, they effectively pass through each other without interference. To calculate the joint effect of both sets of small waves one simply adds, at every point in space and time, the effects of the first one to the effects of the second. The behavior of each set is thus independent of the behavior of the other one. For waves of large amplitude this is usually not true. For large amplitudes, the laws of physics show that the proportionality breaks down. The breaking of a wave is a highly nonlinear process: Once the amplitude is above a certain threshold, the wave behaves in a distinctly new way. It is not just "more of the same" but something new. Nonlinear behavior is common in real life, especially in love and war. As the song says: "Kissing her once ain't half as nice as kissing her twice."

If a system is nonlinear, it is usually much more difficult to understand mathematically than if it were linear. It can also behave in more complicated ways. All this makes the prediction of interacting sets of neurons difficult, especially as the results often turn out to be counterintuitive.

One of the most important technical developments of the last fifty years has been the high-speed digital computer, sometimes referred to as the von Neumann computer after the outstanding mathematician who helped bring it to birth. Because computers, like the brain, can manipulate symbols as well as numbers, it is natural to imagine that the brain is some rather complicated form of von Neumann computer. This parallel, if carried too far, leads to unrealistic theories.

A computer is built with intrinsically fast components. Even in a PC, the basic cycle, or clock rate, is more than 10 million operations per second. For a neuron, on the other hand, a typical firing rate is in the region of only a *hundred* spikes per second. The computer is getting on for a million times faster. A high-speed super-computer, like the Cray, is even faster. Largely speaking, the operations in a computer are serial—that is, one after another. The arrangements in the brain, on the other hand, are usually massively parallel. For example, about a million axons go from each eye to the brain, *all working simultaneously*. This high degree of parallelism is repeated at almost every stage of the system. This kind of wiring compensates somewhat for the relatively slow behavior of neurons. It also means that the loss of a few scattered neurons is unlikely to appreciably alter the brain's behavior. In technical jargon, the brain is said to "degrade gracefully." A computer is brittle. Even a little damage to it, or a small error in its program, may cause havoc. A computer can degrade catastrophically.

A working computer is highly reliable. Given the same input it will usually produce exactly the same output, since its individual components are very reliable. Individual neurons, on the other hand, are much more variable. They are subjected to signals that can modulate their behavior, and some of their properties change while their "computations" are taking place.

A typical neuron can have anywhere from a few hundred to many tens of thousands of inputs and its axon projects as multitudinously. A transistor—a basic unit in a computer—has only a few inputs and outputs.

A computer controlled by a very fast clock can send very accurate messages from one place to another. Each message is usually sent out in parallel along a small number of lines. Along each line the message is binary; that is, either 0 or 1. In most cases some of the lines code for the address; others for the content of the message. Thus, information can be put into one particular place in the computer's memory, quite distinct from other places and, at some later stage, be accessed for further use. Nothing as precise as this occurs in the brain, at least on a small scale.* Thus, almost inevitably, memory in the brain has to be "stored" in a different way.

A brain does not look even a little bit like a general-purpose computer. Different parts of the brain, even different parts of the neocortex, specialize, at least to some extent, in handling different sorts of information. Most memory appears to be stored in the very same locations that carry out current operations. All this is very different from the classical von Neumann computer since, in the computer, the basic computing operations (addition, multiplication, etc.) occur in one or a few places, whereas its memories are stored in many, quite different places.

Finally, while a computer has been deliberately designed by engineers, the brain has evolved by natural selection over many, many generations of animals. This tends to produce a radically different style of design, as outlined in Chapter 1.

It is customary to talk about a computer in terms of hardware and software. Since the person writing the software—the computer's programs—need know very little about the exact details of the hardware (its circuitry, etc.), it has been argued, especially by psychologists, that it is unnecessary to know anything about the "hardware" of the brain.

---

*Charles Anderson and David Van Essen[1] have suggested that the brain has devices to route information from one place to another, but this idea is still controversial.

There is no clear distinction in the brain between hardware and software, and the attempts to foist such theories onto its operations have been unfortunate. The justification for such an approach is that, although the brain is highly parallel, it has some sort of serial mechanism (controlled by attention) on top of all the parallel operations, so that it may superficially appear to be somewhat like a computer at the higher levels of its operations—those far from the sensory inputs.

One can judge a theoretical approach by its fruits. Computers have been programmed so that they do very well solving certain kinds of problems, such as extensive number-crunching, rigid logic, and playing chess. These are things that most people can't do as well or as fast. But faced with tasks that ordinary humans can do in a rapid and effortless way, such as seeing objects and understanding their significance, even the most modern computers fail.

In the last few years there has been considerable effort to design a new generation of computers that act in a more parallel manner. Most of these designs take many small computers, or certain elements of small computers, and connect them so that they all operate simultaneously. Rather complex arrangements handle the exchange of information between the subcomputers as well as the overall control of the computation. These supercomputers have proved to be especially valuable for computations in which the same basic elements of the problem occur in many places, such as in weather prediction.

There have also been moves in the Artificial Intelligence (A.I.) community to create programs that are more brainlike. They substitute a kind of fuzzy logic for the rigid logic usually used in computations. Statements do not have to be true or false, merely more or less probable. The program tries to find the combination of a set of statements that has the greatest degree of probability, and to settle for that conclusion, as against rival conclusions that it assesses to be less probable.[2]

This approach is certainly more brainlike in its conceptual arrangements than earlier A.I. approaches, but it is less like the brain in other ways, particularly in how it stores memory. For this reason it may be difficult to check its agreement with the behavior of real brains at all levels.

A more brainlike approach has been followed by a previously rather obscure group of theorists. This is now called the PDP approach (for Parallel Distributed Processing). The subject has a long history that I will only sketch in outline. One of the earliest attempts was by Warren McCulloch and Walter Pitts in 1943, who showed that "networks" made of very simple units connected together can, in principle, compute any logical or arithmetical function.[3] These networks are now often called "neural networks," because their units are somewhat like very simplified neurons.

This achievement was so impressive that it misled many into believing that the brain worked in this way. Although it probably helped in the design of modern computers, its striking conclusion was highly misleading as far as the brain was concerned.

The next major step was taken by Frank Rosenblatt, who invented a very simple single-layer device that he called a Perceptron. What was significant about this machine was that although its connections were initially random, it could alter them (using a rather simple and explicit rule) so that it could be taught to perform certain simple tasks, such as recognizing printed letters in fixed positions. The way it worked was that the Perceptron could respond to the task in only two ways: true or false. All it had to be told was whether its (tentative) answer was correct or not. It then altered its connections according to a certain rule, called the Perceptron Learning Rule. Rosenblatt proved that for a certain class of simple problems, those called "linearly separable," it could learn to behave correctly after a finite number of trials.[4]

This result attracted a lot of attention because of its mathematical elegance. Unfortunately, its influence was damped by Marvin Minsky and Seymour Papert, who proved that the Perceptron architecture and learning rule could not execute the "exclusive OR" (i.e., apples OR oranges, but not both) and therefore could not learn it. They wrote a whole book expanding on the Perceptron's limitations.[5] This killed interest in Perceptrons for a number of years (Minsky admitted later that it was overkill) and most theoretical work during that time focused on A.I. approaches.*

It is possible to construct multilayer networks of simple units that could easily execute the exclusive OR (or similar tasks) impossible for a simple one-layer Perceptron. Such networks necessarily have many connections at several different levels. The problem is to know which of the originally random connections to alter so that the network performs the operation required of it. Minsky and Papert would have con-

---

*Nevertheless, a number of theorists continued working quietly in the undergrowth, among them Stephen Grossberg, Jim Anderson, Teuvo Kohonen, and David Willshaw.

tributed more if they had produced a solution to this problem rather than beating the Perceptron to death.

The next development to attract widespread attention came from John Hopfield, a Caltech physicist turned molecular biologist turned brain theorist. In 1982, he proposed a network now known as a Hopfield net-work[6] (see Fig. 53). This was a simple network that fed back on itself. Each of its units could have only two outputs: -1 (inhibition) or +1 (excitation). But each unit had a number of inputs. Each such connection had been given a particular strength. The unit added up all the effects* coming at that moment from all its connections. It assumed output state +1 if this sum was greater than zero (i.e., it put out excitation if, on balance, it was more excited than inhibited), otherwise its output was -1. Sometimes this meant that the unit changed its output, because some of the other units had changed their inputs to it.

The calculations were done over and over again until the output of all the units became stable.[†] In Hopfield's network, the adjustment to the state of all the units was not made simultaneously, but one at a time, preferably in a random order. Hopfield was able to show theoretically that, given a set of weights (the strengths of the connections) and any input, the network would neither wander indefinitely nor oscillate but would quickly reach a stable state.[‡]

Hopfield's arguments were cogent, and forcibly and clearly written. His network had an enormous appeal to mathematicians and physicists, who felt at last that here was an idea about the brain to which they could relate (as we say in California). The fact that the network was grossly unbiological in many of its details did not worry them at all.

How is the strength of all the connections adjusted? In 1949, the Canadian psychologist Donald Hebb published a book called *Organization of Behavior*.[7] It was widely believed then (as it is now) that, in the process of learning, one of the key factors is the modification of the strength of the neuronal connections—the synapses. Hebb

---

*The effect of each input on the unit was calculated by multiplying the current input signal (which was either +1 or -1) by the value of the relevant weight. (If the current signal was -1 and the weight was +2, the effect on the unit was -2.)

[†]This network was based on earlier networks that were inspired by a theoretical concept, invented by physicists, called a "spin glass."

[‡]This corresponded to a (local) minimum of a well-defined mathematical function, called an "energy function" because of the spin-glass analogy. Hopfield also gave a simple rule for fixing the weights so that a particular pattern of activity in the network corresponded to a minimum of the energy function.

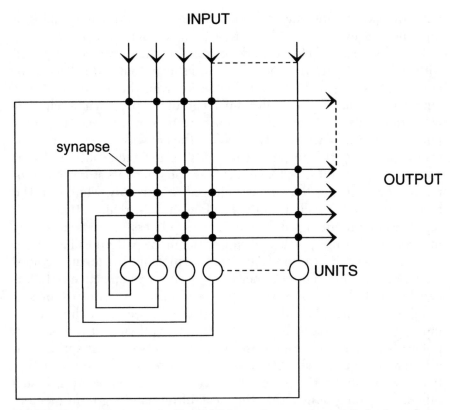

Fig. 53. Wiring diagram of a Hopfield network, sometimes called a "crossbar network." Each small circle represents a "unit," a very oversimplified version of a neuron. The connections are here labelled "synapses." It is the strength of these connections that is adjusted so that the net holds a particular memory.

realized that it was not sufficient that a synapse should be strengthened merely because it was active. He wanted a mechanism that only worked when *two* activities were associated. In a much-quoted passage he said: "When an axon of cell A is near enough to excite a cell B and repeatedly and persistently takes part in firing it, some growth process or metabolic change takes place in one or both cells such that A's efficiency, as one of the cells firing B, is increased." Such a mechanism, or one somewhat resembling it, is now called "Hebbian."

Hopfield used a version of Hebb's rule to adjust the strength of the connections in his network. If, for the pattern in question, two units had the same output, the weights of the two reciprocal connections between them are each made equal to $+1$. If they have opposite outputs, the two weights are each made equal to -1. Loosely speaking, a unit encourages its "friends" and tries to discourage its "enemies."

How does a Hopfield network perform? If the network is supplied with the correct pattern of activity, it will stay in that state. This is not remarkable since in this case it has been given the answer. What is remarkable is that if given only a small part of the pattern as a "clue," it will, after running around a few times, settle down to the right output—that is, to the whole pattern. What the network discovers, by constantly adjusting the outputs of its units, is a stable coalition of unit activities. As a result it will, in effect, have produced its stored "memory" from something that merely nudged its memory. Moreover, the memory is what is called "content addressable"—that is, there is no separate and unique signal that acts as the "address," as there usually is in a computer. Any appreciable part of the input pattern will act as an address. This begins to have some faint resemblance to human memory.

Notice that the "memory" need not be stored in an active state but can be entirely passive, since it is embedded in the pattern of weights—the strength of the connections between all the various units. The network can be completely inactive (with all the outputs put to 0), yet when a signal is fed in, the network will spring into action and in quite a short time will settle into a state of steady activity corresponding to the pattern that had to be remembered. It is surmised, on good grounds, that the recall of human long-term memory has this general character (although the pattern of activity does not last indefinitely). You are capable of remembering a vast number of things that you are not remembering at this moment.

Neural networks (and Hopfield networks in particular) can "remember" one pattern, but can the same network also remember a second pattern in addition to the first? If the patterns are not too similar, a network can remember several distinct patterns—that is, given a sufficient part of any one of these patterns the network will cycle a few times and then produce that particular pattern as its output. In these systems the memory is distributed, because any one memory is distributed over many connections. The memories are superimposed, because any one connection can be involved in several memories. In addition, the memory is robust, since altering a few connections will usually not alter its behavior very much.

Not surprisingly there is a price to pay for these properties. If too many memories are added to the network it can easily become confused. Given a clue or even a complete pattern as input, it may produce a nonsense output.* It has been suggested[8,9] that this is what happens in our dreams (a process Freud called "condensation") but that is another story. It should be noticed that all these properties are "emergent." They were not deliberately crafted by the designer of the network, but arose from the nature of the units, their pattern of connections, and the rules for adjusting the weights.

Another property of a Hopfield network is that if the weights of its connections have been calculated as appropriate for several inputs that *do* largely resemble each other, it will "remember" a sort of average of what it has been trained on. This is another somewhat brainlike property. In humans we hear a particular vowel as the same even if the relevant sound varies over a certain range. The inputs are different, but similar. The output—what we hear—is the same.

These simple networks are far removed from the complexities of the brain, but their simplicity does allow us to understand their behavior. The properties that emerge from even simple networks may well emerge from more complex networks having the same general character. Moreover, they provide us with sets of ideas that shed light on what particular brain circuits may be doing. For example, the connections of a region of the hippocampus called CA3 do in fact look like a content-addressable network. Whether this is correct or not has, of course, to be verified by experimentation.

Interestingly, these simple neural networks have some of the properties of a hologram. In a hologram, several images can be stored on top of one another; any part of a hologram can be used to regenerate the whole image, although with reduced clarity; the holograph is robust to small imperfections. This analogy between the brain and a hologram has often been enthusiastically embraced by those who know rather little about either subject. It is almost certainly unrewarding, for two reasons. A detailed mathematical analysis has shown that neural networks and holograms are mathematically distinct.[10] More to the point, although neural networks are built from units that have some resemblance to real neurons, there is no trace in the brain of the apparatus or processes required for holograms.[†]

---

*For a Hopfield network this can be shown to be related to a weighted sum of those of its stored "memories" that are, in a precise sense, closely related to the output.

†The hologram led Christopher Longuet-Higgins to invent in 1968 a device he called a "holophone." This led him to another device called a "correlogram" that then turned into a particular type of neural network. This was studied in detail by his student David Willshaw for his Ph.D. thesis.

*   *   *

A more recent publication that produced a dramatic impact was the book *Parallel Distributed Processing*, a thick two-volumed work by David Rumelhart, James McClelland, and the PDP group.[11] This appeared in 1986 and rapidly became a best-seller, at least in academic circles. I am nominally a member of the PDP group, and with Chiko Asanuma contributed a chapter to the book, but my role might be better described as that of a fringee, or perhaps a gadfly. Almost my only contribution to their efforts was to insist that they stop using the word *neurons* for the units of their networks.

The Psychology Department of the University of California, San Diego, is only a mile or so from the Salk Institute. In the late seventies and early eighties I used to walk to the small, informal meetings of their discussion group. Now the land over which I strolled has been turned into enormous parking lots. The pace of life has quickened so that I now dash there and back by car.

At that time the group was led by David Rumelhart and Jay McClelland, although before long Jay left for the East Coast. Both were mainly psychologists in origin, but they had become dissatisfied with symbol-processing machines and together had developed an "interactive activator" model of word processing. Encouraged by Geoffrey Hinton, another student of Christopher Longuet-Higgins, they embarked on a more ambitious "connectionist" program. The term *parallel distributed processing* was adopted because it covered a wider field than the earlier term *associative memory*.*

When networks were first invented, some theorists heroically soldered together small yet clumsy electrical circuits, often containing old-fashioned relays, in order to simulate their very simple networks. The recent development of more complex neural networks has been greatly helped by the much increased speed and cheapness of modern digital computers. New ideas about networks can now be tested by simulations in a computer (often a digital computer), rather than solely by crude analogue models or by rather difficult mathematical arguments, as was earlier the case.

The 1986 PDP book was a long time in gestation, having started in late 1981. This was fortunate because it was the late development of a

---

*Their interactions with some other like-minded theorists had produced a previous book, published in 1981, called *Parallel Modes of Associate Memory*, edited by Geoffrey Hinton and Jim Anderson. While this was read by people working on neural networks, it did not have the wide impact of the later book.

particular algorithm (or, rather, its rediscovery and exploitation) that, on top of the earlier work, made such an immediate impression. Their book was read eagerly not only by brain theorists and psychologists but by mathematicians, physicists, engineers, and even by people working in Artificial Intelligence, though in the latter case their reaction was initially rather hostile. News of it eventually penetrated to neuroscientists and molecular biologists.

Subtitled "Explorations in the Microstructures of Cognition," it is something of a mixed bag, but one particular algorithm in it produced striking results. This algorithm is known as "the backpropagation of errors," but it is usually simply referred to as backpropagation, or "backprop" for short. To understand it you need to know a bit about the general nature of learning algorithms.

Some forms of learning in neural networks are called "unsupervised." This means that no teaching information is fed in from outside. The change made to any particular connection depends only upon what is happening locally within the system. A simple Hebbian rule has this character. In supervised learning, on the other hand, teaching information about the performance of the networks is provided to the network from outside the immediate system.

Unsupervised learning has a powerful appeal because, in a sense, the network teaches itself. Theorists have devised more powerful learning rules, but these need a "teacher" to tell the network whether it made a good, poor, or bad response to some inputs. One such rule is called the "delta rule."

To train a network it is necessary to have a set of inputs to train it on. (We shall see an example of this shortly when we consider NETtalk.) This is known as the "training set." To be useful it must be a fair sample of the inputs the network is likely to encounter after it has been trained. Usually, the signals of the training set must be fed in several times, so that a lot of training is often required before the network has learned to perform well. This is partly because such networks tend to be set up with random connections. The initial connections in the brain, being controlled to some extent by genetic mechanisms, are usually not completely random.

How is the network trained? A signal from the training set is fed into the network and the network produces an output. This means that the output neurons are each in a special state of activity. The teacher then signals to each output unit what its error is—that is, how its activity differs from the correct one. This difference—between the real activity and the desired one—is the origin of the term *delta*, often used in mathematics for a small but finite difference. The algorithm, the network's learning rule, then uses this information to calculate how the weights should be altered to improve the network's performance.

One of the earlier teaching rules was presented in 1960 for a network called the "Adaline," invented by Bernard Widrow and M. E. Hoff, so that the delta rule is also called the Widrow-Hoff rule. The rule is designed so that at every step the total error is always reduced.* This means that, as training proceeds, the network eventually reaches an error minimum. This much is certain. What is uncertain is whether this is the true global minimum or merely a local one. In terms of a landscape, have we reached a lake in a volcano, a lower-lying pond, the ocean, or a sunken sea like the Dead Sea?

The training algorithm can be adjusted so that the steps toward a local minimum can be large or small. If large, the algorithm may make the network jump all around the local minimum (it will start to travel downhill but will go too far and so go up again). If small, it may take an excessively long time to reach the bottom of the minimum. More elaborate schemes of adjustment can also be used.

Backpropagation is a special example of a learning algorithm requiring a teacher. For it to work the units of the network have to have a special feature. They must not produce a binary output (either 1 or 0; alternatively +1 or -1) but a graded one. This is usually made to lie *between* 0 and 1. It is fondly believed by theorists that this corresponds to the average rate of firing of a neuron (the maximum rate being taken as 1), but they are usually vague about the time over which this average should be taken.

How is it decided what this "graded" output should be? As before, each unit takes a weighted sum of its inputs, but this time there is no true threshold. If the sum is very small, the output is almost zero. If a bit larger, the output increases, and if the sum is very large the output approaches its maximum. The typical relationship between the summed input and the output is the sort of sigmoid curve shown in Figure 54. This is not too different from the behavior of real neurons, with the average firing rate taken as the output.

This harmless-looking curve has two important properties. Mathematically it is "differentiable," meaning here that it has a finite

---

*More precisely, the mean square error is reduced, so this rule is sometimes called the LMS (Least Mean Square) rule.

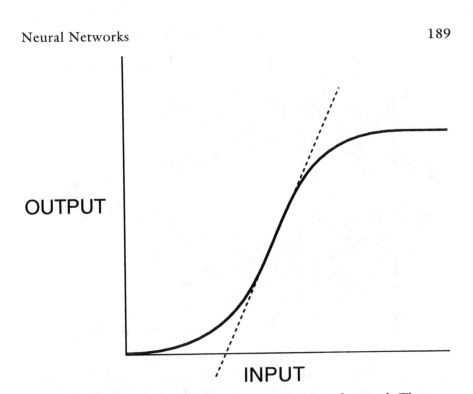

Fig. 54. A typical input-output curve for a unit in a neural network. The curve is nonlinear. (The dotted line shows an example of a linear curve.)

slope everywhere; the backprop algorithm depends on this property. More important, the curve is nonlinear, as indeed real neurons are. Doubling the (internal) input does not necessarily produce double the output. This nonlinearity enables it to handle a much larger range of problems than can be handled by a strictly linear system.

Now let's look at a typical backprop network. There are usually three distinct layers of units (see Figure 55). The lowest layer is the input layer. The next is called the "hidden units" layer, because the units do not connect directly with the world outside the network. The top layer is the output layer. In the lowest layer each unit connects to every other unit in the layer immediately above. The same is true for the middle layer. The network has only forward connections. It has no sideways connections and no back projections (except for training it). Its construction could hardly be simpler.

When the training starts, all the weights are simply given values at random, so the network's initial response to any signal is largely nonsensical. It is then given one of the training inputs, produces an out-

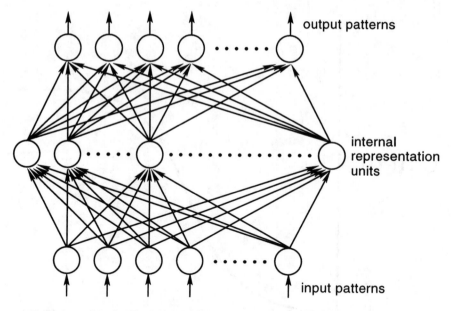

Fig. 55. A simple multilayer network. Each unit connects to all units in the layer above it. There are no sideways connections, or back connections. The "internal representation units" are often referred to as the "hidden units."

put and adjusts its weights according to the backprop training algorithm. This is done as follows: Each unit in the upper layers is told (after the network has produced its output to a training input) how that output differs from the "correct" output. It uses this information to make small adjustments to the weights of the synapses made onto it from each of the units in the layer below. It then backpropagates this information to each of the units in the hidden layer. Each hidden unit then collects this error information sent to it by all the upper layer units, and uses this information to adjust all the synapses coming into it from the lowest layer.

The details of the algorithm are such that the network, as a whole, is always adjusted to reduce its errors. This process is then repeated many times. (The algorithm is sufficiently general that it can be applied to feed-forward networks having more than just three layers.)

After a sufficient amount of training the network is ready for use. It can then be tested on a "test set" of inputs, chosen to be similar in its general (statistical) properties to the training set but otherwise distinct from it. (At this stage the weights are kept constant in order to see how the trained network performs.) If the performance is not satisfactory the designer usually starts all over again, modifying the architecture, the way the input and output are coded, the parameters of the training rule or the amount of training.

All this sounds rather abstract, but an example may make it clearer. One of the most striking demonstrations was provided by Terry Sejnowski and Charles Rosenberg in 1987.[12] The task they set their network, which they called NETtalk, was to produce spoken English from a written English text. This is not straightforward since English is an especially difficult language to pronounce because of its irregular spelling. The network was, of course, not explicitly given any of the rules of English pronunciation. It had to learn them just from the corrections it received after each of its attempts during the training sessions. The input was fed into the network, letter by letter, in a special way. The overall output of NETtalk was a string of symbols related to spoken sounds. To make the demonstration more vivid, this output was coupled to a quite separate, preexisting machine (a digital speech synthesizer) that produced speech sounds from the output of NETtalk, so that it was possible to listen to the machine "reading" the English text.

Since the pronunciation of an English letter depends largely on what letters lie before and after it, the input layer looked at strings of seven letters at a time.* The output layer had a unit for each of twenty-one "articulatory features"[†] of the required phonemes plus five units to handle syllable boundaries and stresses. The general architecture[‡] is shown in Figure 56.

They used extracts from two texts to train the network, both of which were supplied with the phonetic transcriptions needed to train the machine. The first text was an excerpt of *Merriam-Webster's Pocket Dictionary*. Their second choice, somewhat surprisingly, was the continuous speech of a child. The weights initially had small random values and were updated after every word during the training period. The program was written so that the computer did this automatically from

---

*There had to be a unit for each of twenty-nine "letters"; twenty-six for the letters of the alphabet plus three for punctuation and word boundaries. Thus there had to be $29 \times 7 = 203$ units in the input layer.

[†]As an example, the consonants p and b are both called "labial stops" because they start with pursed lips.

[‡]The original middle (hidden) layer had eighty hidden units, but a later one, with 120, performed somewhat better. There were about twenty thousand synapses whose weights the machine had to adjust. The weights could be positive or negative. They did not construct a real, highly parallel network to do all this, but simulated the network on what was then a moderately fast computer (a VAX 11/780 FPA).

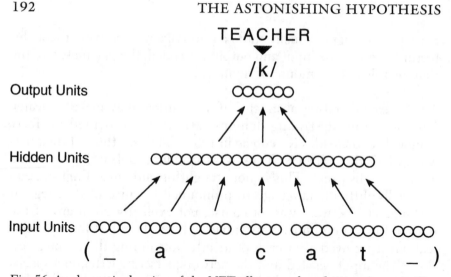

Fig. 56. A schematic drawing of the NETtalk network architecture, a specific example of the general scheme in Figure 5.5. A window of seven "letters" in an English text ("a cat" here) is fed to an array of 203 input units. Information from these units is transformed by an intermediate layer of 80 "hidden" units to produce patterns of activity in 26 output units.

the input and (correct) output information provided. In judging the real output the program accepted the phoneme nearest the true one as the best guess. This often had contributions from several of the "articulatory" output units.

It is fascinating to listen to the machine learning to "read" English.* At first, because of the initially random connections, only a confused string of sounds is heard. NETtalk soon learns the distinction between vowels and consonants, but at first it only knows one vowel and one consonant, so it appears to babble. Next it recognizes word boundaries, and produces strings of pseudo-words. After about ten passes through the training set the words become intelligible and sound very similar to a small child talking.

Naturally the performance was not perfect. In some cases English pronunciation depends on meaning, of which NETtalk knows nothing. A common confusion was between rather similar sounds, such as the "th" sound in "thesis" and "throw." Tested on a further sample of text

---

*The computer used could not work fast enough to produce its output in real time, so the output had to be recorded and then speeded up so that one could understand it.

from the same child it did quite well, showing that it could transfer its learning from the rather small training set (1,024 words) to novel words it had not encountered before.* This is called "generalization."

Clearly the network was not merely a look-up table for each word it had been trained on. Its powers of generalization depend on the redundancy of English pronunciation. Not every English word is pronounced in its own unique way, although foreign speakers tackling English for the first time are apt to think so. (This problem springs from the dual origin of English from both the Romance and Germanic languages, an origin that helps make the vocabulary of English so rich.)

One advantage of a neural network over most collections of real neurons is that it is easy, after training, to examine each hidden unit to see what its receptive field is. Would one letter activate just a few hidden units, or would the activity be more holographic and spread over many hidden units? The answer was somewhat closer to the former. There was no special hidden unit for every letter-to-sound correspondence, but nevertheless one such correspondence was not spread widely over all the hidden units.

Because of this they were able to test how the behavior of the hidden units clustered (i.e., had properties in common). Sejnowski and Rosenberg found that ". . . the most important distinction was the complete separation of consonants and vowels. However within these two groups the clustering [of the hidden units] had a different pattern. For the vowels, the next most important variable was the letter, whereas consonants were clustered according to a mixed strategy that was based more on the similarity of their sounds."

The importance of this rather messy arrangement, which is typical of neural networks, is that it is strikingly similar to the way many real cortical neurons respond (e.g., in the visual system), and quite different from the neat design an engineer might impose on the system.

They concluded:

NETtalk is an illustration in miniature of many aspects of learning. First, the network starts out with considerable "innate" knowledge in

---

*Sejnowski and Rosenberg also showed that the network was fairly resistant to random damage they gave to its connections, and its performance "degraded gracefully" under these circumstances. They also experimented with groups of eleven input letters (instead of seven). This improved the performance significantly. Adding a second hidden layer did not improve performance but the network did better at generalization.

the form of input and output representations that were chosen by the experimenters, but with no knowledge specific for English—the network could have been trained on any language with the same set of letters and phonemes. Second, the network acquired its competence through practice, went through several distinct stages, and reached a significant level of performance. Finally, the information was distributed in the network such that no single unit or link was essential. As a consequence, the network was fault tolerant and degraded gracefully with increasing damage. Moreover, the network recovered from damage much more quickly than it took to learn initially.

Despite these similarities with human learning and memory, NETtalk is too simple to serve as a good model for the acquisition of reading skills in humans. The network attempts to accomplish in one stage what occurs in two stages of human development. Children learn to talk first, and only after representations for words and their meanings are well developed do they learn to read. It is also very likely that we have access to articulatory representations for whole words, in addition to our ability to use letter-to-sound correspondences, but there are no word level representations in the network.

Notice that nowhere in the network are the rules of English pronunciation laid out explicitly as they would be in a standard computer program. They are embedded *implicitly* in the learned pattern of the weights. This is just how a child learns a language. It learns to speak correctly but has no idea of the tacit rules its brain is using to do so.*

Several features of NETtalk are quite unbiological. The units violate the rule that one neuron produces only excitation or only inhibition but not both. More seriously, the backprop algorithm, taken literally, requires that teaching information be rapidly sent back along exactly the same axons that carried the forward, operational information. This is most unlikely to happen in the brain. Attempts have been made to produce separate circuits to do this, but to me they appear very forced and unbiological.

---

*NETtalk has many simplifications in addition to those outlined above. Although the authors believe in distributed representations, they have "grandmothered" both the input and the output—that is, a single unit stands for, say, "the letter *a* in the third position in the window." This is done to reduce the time needed for computation, and is a reasonable form of simplification. Then the way the data are fed into the seven-letter window in successive steps seems quite unbiological, although perfectly acceptable for an A.I. program. The winner-take-all step at the output is also not implemented by "units," nor is a set of units provided to express the difference between the desired output and the actual output—the teaching signal. Both these operations are performed by the program.

In spite of all these limitations, NETtalk is a very impressive demonstration of what a relatively simple neural network can do. Recall that it has less than five hundred units and twenty thousand connections. The numbers would be higher if some of the limitations and omissions (listed above in a footnote) were included, but this would probably not increase the numbers by more than, say, a factor of ten. There are about five thousand neurons in the neocortex beneath a patch of surface about a quarter of a millimeter on each side (less than the size of a pinhead), so compared to the human brain as a whole, NETtalk is minute.* It is all the more impressive that it can learn such a relatively complex task.

Another neural network was produced by Sidney Lehky and Terry Sejnowski.[13] The problem they set their network was to try to deduce the 3D shape of certain objects from their shading (the so-called shape-from-shading problem described in Chapter 4) without the network being told the direction of the light source. The surprise came when the receptive fields of the units of the hidden layer were examined. Some of these were very similar to those of some of the neurons found experimentally in the brain's first visual area, V1. These had always been taken to be edge detectors or bar detectors, but the network, during its training, had been shown neither edges nor bars. Nor had the network's designers imposed the receptive fields. They emerged as a result of the training. In addition, when the network was tested on a bar, the output layer units responded like the complex cells with end-stopping in V1 (described on page 144).

Both the network and the backprop algorithm were unbiological in several ways but this example raises a point that, in retrospect, should have been obvious: *One cannot deduce the function of a neuron in the brain merely by looking at its receptive field.* As mentioned in Chapter 11, it is also important to know its projective field—that is, what neurons it sends its axon to.

We have focused upon two extreme cases of "learning" in neural networks, unsupervised learning (as exemplified by Hebb's rule) and supervised learning (as in backprop), but there are several other general classes. An equally important one is known as "competitive learn-

---

*This comparison is not quite fair because a single unit in a neural network is better thought of as equivalent to a small group of related neurons in the brain, so perhaps eighty thousand neurons (the number beneath about one square millimeter) would be a more appropriate figure.

ing."* The basic idea is that in the operation of the network there is a winner-take-all mechanism so that one unit (or more realistically, a few of them) suppresses all the other units, since it best expresses the significance of the input. Instead of adjusting all the connections in the system during the learning process, only those closely associated with that winner are adjusted for that step. This is usually modeled using a three-layer network (like the standard backprop network) but with the crucial difference that there are strong interconnections between the units in the middle layer. The strength of these connections is usually fixed, rather than modifiable. They are often arranged to be excitatory at short range but inhibitory at long range; a unit tends to make friends with its immediate neighbors but antagonizes those farther away. This arrangement means that the units in the middle layer are competing for the activity of the entire network. In a carefully designed network there is often just a single winner on any one trial.

For such a network there is no external teacher at all. The network itself finds the best response. The learning algorithm is such that only the winning unit and its neighbors adjust their incoming weights. They do so in such a manner that that particular response is now more likely to be made in the future. Each hidden unit learns to associate itself with one particular sort of input, since the learning algorithm automatically pushes the weights in the required direction.†

All the networks we have considered so far have dealt with static inputs that produced, after an interval, a static output. Clearly there are brain operations that express a time sequence, such as whistling a tune or understanding and speaking a language. Networks have been designed to tackle these problems in a primitive way but they have not as yet been developed very far. (It is true that NETtalk produces a time-sequence, but this is not a property of the network but of the way the data are fed into and out of the network.)

As linguists have emphasized, the operations of speech (such as the rules of syntax) are at the moment better handled by programs written according to A.I. theory. At bottom this is because networks are best at

---

*Developed by Stephen Grossberg, Teuvo Kohonen, and others.
†I will not enter into the limitations of competitive networks. Obviously there must be enough hidden units to accommodate everything the network is trying to teach itself from all the inputs provided. The training must not be too fast or too slow, and so on. Such networks usually require careful design to work properly. No doubt more complex applications of the basic idea of competitive learning will be invented in the near future.

highly parallel processing, whereas these linguistic tasks demand a degree of serial processing. The brain has attentional systems, having a somewhat serial nature, that operate on the underlying parallel processing. So far neural networks have not reached the degree of complexity that demands such serial processes, although this is bound to come.

Real neurons (their axons, their synapses, and their dendrites) experience unavoidable time delays and change during their processing time. The designers of most neural networks have regarded these properties as nuisances to be avoided. This is probably the wrong attitude. Evolution will almost certainly have built on these changes and these time delays and used them to advantage.

A possible criticism of these neural networks is that since they use such a grossly unrealistic learning algorithm, they really don't reveal much about the brain. There are two counters to this. One is to try algorithms that appear to be biologically more acceptable. The other is a more general and powerful one. David Zipser, a molecular biologist turned neural theorist (now at the University of California, San Diego), has pointed out that backprop is really a very good way of identifying the nature of the system under study.[14] He calls this "neural system identification." The claim is that if the architecture of the network at least approximates the real thing, and if enough constraints on the system are known, then backprop, being a method that minimizes errors, will usually arrive at a solution that, *in its general characteristics*, will resemble the real biological one. It will thus provide a first step in the right direction toward understanding how the biological system is behaving.

If the neurons and the architecture of their connections are tolerably realistic, and if enough constraints have been imposed on the system, the model produced may be sufficiently like the real thing to be useful. This would allow the behavior of the components of the model to be examined in detail. This can be done more rapidly and thoroughly than the equivalent experiments on an animal.

It is important to realize that the scientific task does not end there. The model may show, for example, that a certain class of synapses in the model have to be alterable in a certain way, dictated in the model by backprop. But in the real system, backprop does not occur. So the modeler must find the appropriate *realistic* learning rule for that class of synapse. For example, a version of Hebb's rule may be all that is required for those particular synapses. The model must then be rerun with these realistic (local) learning rules (which may be different in different parts of the model) plus any more global signals, if these appear to be involved.

If the model still works, then the experimentalists must show that such types of learning do indeed occur at the predicted places, and support this by uncovering the cellular and molecular mechanisms involved in such learning. Only by these means can we proceed from the play of "interesting" demonstrations to results with the hard ring of true science.

All this means that a lot of models, or variants of them, need to be tested. Fortunately, the development of extremely rapid and cheap computers now allows many models to be simulated, so one can test whether a certain arrangement does actually behave in the way one hopes it does. But even with the latest computers it is difficult to test models that are as extensive and complicated as one might wish.

The insistence that all models should be tested by simulation has had two unfortunate by-products. If the postulated model behaves fairly successfully, its author often finds it hard to believe that it's not correct. Yet experience has shown that several quite distinct models may produce the same behavior. To decide which of these models approximates best to the truth, it is likely that other evidence will be necessary, such as the exact properties of the real neurons and molecules in that part of the brain.

The other hazard is that an overemphasis on successful simulations may inhibit thinking about the problem in less restrictive ways, and thus stifle theoretical creativity. Nature works in peculiar ways. Too narrow an approach to a problem may lead one to abandon a potentially fruitful idea because of a particular difficulty. Yet evolution may have produced some little additional trick to circumvent that difficulty. In spite of these reservations it is sensible to simulate a theory, if only to get some feel as to how it really works.

What can we conclude about neural networks? Their basic design is more brainlike than the architecture of a standard computer. Nevertheless, their units are not as complex as real neurons, and the architecture of most networks is grossly oversimplified compared to the circuitry of the neocortex. At the moment, networks have to be extremely small if they are to be simulated in a reasonable time on standard computers. This will change as computers get faster, and as highly parallel (network-like) computers are produced commercially, but it is always likely to be a serious handicap.

In spite of these limitations, neural networks now show an astonishing range of performance. The whole field is bubbling with new ideas. Although many of these will probably fade into oblivion, there will be steady progress in understanding them, grasping their limitations and devising new tricks to improve them. The fact that such networks are likely to have important commercial applications, even though these sometimes lead theorists too far away from biological reality, will in the long run produce both useful ideas and useful gadgetry. It may turn out that the most important result of all this work on neural networks is that it suggests new ideas about how the brain might operate.

In the past, many aspects of the brain seemed completely incomprehensible. Thanks to all these new concepts one can now at least glimpse the possibility that some day it will be possible to model the brain in a biologically realistic way, as opposed to producing biologically implausible models that only capture somewhat limited aspects of brain behavior. Even now these new ideas have sharpened our approach to experimentation. We now know more about what we need to understand about individual neurons. We can point to aspects of the circuitry that we don't grasp well enough (such as the back pathways in the neocortex). We see the behavior of single neurons in a new light, and realize that the behavior of whole groups of them is the next important task on the experimental agenda. Neural networks have a long way to go but they have, at last, got off to a good start.

PART III

14

Visual Awareness

"Philosophy is written in that great book that lies before our gaze—I mean the universe—but we cannot understand it if we do not first learn the language and grasp the symbols in which it is written."

—Galileo

Let us now survey the ground we have covered so far. The main theme of the book is the Astonishing Hypothesis—that each of us is the behavior of a vast, interacting set of neurons. Christof Koch and I suggested that the best way to approach the problem of consciousness was to study visual awareness, both in man and his near relations. However, the way people see things is not straightforward. Seeing is both a constructive and a complicated process. Psychological tests suggest that it is highly parallel but with a serial, "attentional" mechanism on top of the parallel one. Psychologists have produced several theories that try to explain the general nature of the visual processes, but none is much concerned with how the neurons in our brains behave.

The brain itself is made of neurons (plus various supporting cells). Each neuron, considered in molecular terms, is a complex object, often with a rather bizarre, irregular shape. Neurons are electrical signallers. They respond quickly to incoming electrical and chemical signals and dispatch fast electrochemical pulses down their axon, often over distances very many times greater than the diameter of their cell bodies. There are enormous numbers of them, of many distinct types, and they interact with each other in complicated ways.

The brain is not a general-purpose machine, like most modern computers. Each part, when fully developed, does a somewhat different and specific job, but, in almost any response, many parts interact together. This general picture is supported by studies of humans whose brains have been damaged and by modern methods of scanning the human brain from outside the head.

The visual system has far more distinct cortical areas than one might have expected. These areas are connected in a semihierarchical manner. A neuron in one of the lower cortical areas (i.e., those connected most closely to the eyes) is mainly interested in relatively simple aspects of only a small fragment of the visual scene, although even these neurons are influenced by the visual context of the fragment. The ones at the higher cortical levels respond best to more complex visual objects (such as faces or hands) and are not too fussy about exactly where these objects are in the visual scene. There appears to be no single cortical area whose activity corresponds to the global content of our visual awareness.

To understand how the brain works, we have to develop theoretical models that describe how sets of neurons interact with each other. At the moment, the neurons in these models are oversimplified. Modern computers, while many times faster than the ones available a generation ago, can simulate only a relatively small number of these neurons and their interconnections. Nevertheless, these primitive models, of which there are several distinct types, often show surprising behavior, not unlike some of the behavior of the brain. They help provide us with new ways of thinking about how the brain might work.

This is the background, then, against which we have to approach the problem of visual awareness: how to explain what we see in terms of the activity of neurons. In other words, What is the "neural correlate" of visual awareness? Where are these "awareness neurons"—are they in a few places or all over the brain— and do they behave in any special way?

Let's start by looking once again at the ideas briefly outlined in Chapter 2. Exactly what psychological processes are involved in visual awareness? If we can find where these different processes are situated in the brain, this knowledge may help us to locate the awareness neurons we are looking for.

Philip Johnson-Laird suggested that the brain has an operating system—as a modern computer has—and its actions correspond to consciousness. In his book *Mental Models* he puts this idea in a wider context. He suggests that the division between conscious and unconscious processes is a result of the very high degree of parallelism in the brain. Such parallel processing allows the organism to evolve special sensory, cognitive, and motor systems that operate rapidly, since many of their neurons can work at the same time (rather than one after another) as I have already described for the visual system. The overall control of all this activity by the more serial operating system enables decisions to be made rapidly and flexibly. A very rough analogy would be to an orchestral conductor (the operating system) controlling the parallel activities of all the members of an orchestra.

While this operating system can monitor the output of the neural systems it controls, he postulates that it does not have access to the details of their operations but only the results they present to it. By introspection we have access to only a limited amount of what is going on in our brains. We have no access to the many operations that lead up to the information given to the brain's operating system. As he puts it, in introspection, "We tend to force intrinsically parallel notions into a serial straitjacket," since he envisages the operating system as operating largely in a serial manner. This is why introspection can be so misleading.

Johnson-Laird's ideas are clearly and forcibly expressed, but if we want to understand the brain in neural terms we have to identify the location and nature of this operating system. It may not necessarily have exactly the same properties as those found in modern computers. The brain's operating system is probably not cleanly located in one special place. It is more likely to be distributed, in two senses: It may involve separate parts of the brain interacting together, and the active information in one of these parts may be distributed over many neurons. Johnson-Laird's description of the brain's operating system reminds one somewhat of the thalamus, but the neurons of the thalamus may be too few to represent all the contents of visual awareness, although this could be tested. It seems more probable that some (but not all) neurons in the neocortex do this, probably under the influence of the thalamus.

At which stages in the various functional brain hierarchies should we look for the neural correlates of awareness? Johnson-Laird believes that the operating system is at the highest level of the processing hierarchies whereas, as we saw, Ray Jackendoff thinks that consciousness is more closely associated with the intermediate levels. Which idea is the more plausible?

Jackendoff's view* of visual awareness is based on David Marr's idea of the 2½D sketch—roughly the viewer-centered representation of visible surfaces described in Chapter 6—rather than Marr's 3D (three dimensional) model. This is because humans directly experience only the presented side of objects in the visual field; the presence of the invisible rear of an object is only an inference. On the other hand, he believes that visual understanding—what one is aware of—is determined by the 3D model together with "conceptual structures"—fancy words for thoughts. This illustrates what he means by the intermediate-level theory of consciousness.

An example may make this clearer. If you look at a person whose back is turned to you, you can see the back of his head but not his face. Nevertheless, your brain infers that he has a face. We can deduce this because if he turned around and showed that the front of his head had no face, you would be very surprised. The viewer-centered representation corresponds to what you saw of the back of his head. *It is what you are vividly aware of.* What your brain infers about the front would come from some sort of 3D model representation. Jackendoff believes you are not directly conscious of this 3D model (nor of your thoughts, for that matter). Recall the old line: How do I know what I think till I hear what I say.

I have put Jackendoff's penultimate version of his theory in the footnotes,[†] as his wording is not easy to understand on a first reading.[‡] Applied to vision, what he means (if I understand him correctly) is that the "distinctions of form"—that is, the position, shape, color, motion, and so on of a visual object—are related to (caused by/supported by/projected from) a short-term memory representation that results from winner-take-all mechanisms (the "selection function"), and that this representation is "enriched" by attentional processing.

The value of Jackendoff's approach is that it warns us not to assume

---

*It is not easy to summarize Jackendoff's ideas without distorting them. Readers who would like to have a deeper understanding should consult his book. I have not attempted to convey his arguments about phonology, syntax, and linguistic meaning nor his approach to musical cognition. Instead, I have tried to condense his basic ideas, especially as they apply to vision.

[†]In his own words: "The distinctions of form present in each modality of awareness are caused by/supported by/projected from a structure of intermediate level for that modality that is part of the matched set of short-term memory representations designated by the selection function and enriched by attentional processing. Specifically, linguistic awareness is caused by/supported by/projected from phonological structure; musical awareness from the musical surface; visual awareness from the 2½D sketch."

[‡]To appreciate the subtleties of his wording the reader should consult Jackendoff's book. (The final version of his theory, Theory VIII, deals also with Affect.)

that the highest levels of the brain must necessarily be the only ones involved in visual awareness. The vivid representation in our brains of the scene directly before us may involve various intermediate levels. Other levels may be less vivid or, as he suggests, we may not be truly conscious of them at all.

This does not mean that information flows only from the surface representation to the 3D one; it almost certainly flows in both directions. When you imagine the front of the face in the example above, what you are aware of is a conscious surface representation generated by the unconscious 3D model. This distinction between the two types of representation will probably have to be refined as the subject develops, but it gives us a rough first idea of what it is we are trying to explain.

Exactly where these levels may be located in the cortex is not so clear. For vision, they might correspond more to the middle parts of the brain (such as the inferotemporal regions plus certain parietal regions) rather than to the frontal regions of the brain, but exactly which parts of the visual hierarchy (diagrammed in Fig. 52) he is referring to remains to be discovered. (This is discussed more fully in Chapter 16.)

Having seen how some psychologists view the matter, let's now look at the problem from the point of view of a neuroscientist who knows about neurons, their connections, and the way they fire. What is the general character of the behavior of the neurons associated (or not) with consciousness? In other words, what is the "neural correlate" of consciousness? It is plausible that consciousness in some sense requires the activity of neurons. It may be correlated with a special type of activity of some of the neurons in the cortical system. Consciousness can undoubtedly take different forms, depending on which parts of the cortex are involved. Koch and I have hypothesized that there is only one basic mechanism (or a few) underlying them all. We expect that, at any moment in time, consciousness will correspond to a particular type of activity in a transient set of neurons that are a fraction of a much larger set of potential candidates. The questions at the neural level then become:

- Where are these neurons in the brain?
- Are they of any particular neuronal type?
- What is special (if anything) about their connections?
- What is special (if anything) about the way they are firing?

How could we go about finding which neurons are involved in visual awareness? And are there any clues that might suggest the manner of their firing that corresponds to such awareness?

As we have seen, there are several hints from psychological theories. Awareness is likely to involve some form of attention, so we should study the mechanism the brain uses to attend to one visual object rather than another. Awareness is likely to involve some form of very short-term memory, so we should try to discover how neurons behave when storing and using such memories. Finally, we seem to be able to attend to more than one object at a time. This poses problems for some neuronal theories of awareness, so let us deal with this first.

What happens in the brain when we see an object? There are an almost infinite number of possible, different objects that we are capable of seeing. There cannot be a single neuron, often referred to as a "grandmother cell," for each one. The combinatorial possibilities for representing so many objects at all different values of depth, motion, color, orientation, and spatial location are simply too staggering. This does not preclude the existence of sets of somewhat specialized neurons responding to very specific and ecologically highly significant objects like the appearance of a face.

It seems probable that, at any moment, any particular object in the visual field is represented by the firing of a *set* of neurons.* Because any object will have different characteristics (form, color, motion, etc.) that are processed in several different visual areas, it is reasonable to assume that seeing any one object often involves neurons in many different visual areas. The problem of how these neurons temporarily become active as a unit is often described as "the binding problem." As an object seen is often also heard, smelled, or felt, this binding must also occur across different sensory modalities.[†]

Our experience of perceptual unity thus suggests that the brain in some way binds together, in a mutually coherent way, all those neurons actively responding to different aspects of a perceived object. In other words, if you are currently paying attention to a friend discussing some point with you, neurons that respond to the motion of his face, neurons that respond to its hue, neurons in your auditory cortex that respond to the words coming from his face, and possibly the memory traces associ-

---

*This would not cause any special problem if the members of the set were in close proximity (implying that they probably interact somewhat), received somewhat similar inputs, and projected to somewhat similar places. In that case they would be like the "neurons" in one single neural net. Unfortunately, such simple neural nets can usually handle only one object at a time.

[†]It is not completely certain that the binding problem (as I have stated it) is a real one, or whether the brain gets around it by some unknown trick.

ated with knowing whose face it is all have to be "bound" together, to carry a common label identifying them as neurons that jointly generate the perception of that specific face. (The brain can sometimes be tricked into making an incorrect binding, as when you hear the voice coming not from the ventriloquist but from his dummy.)

There are several types of binding. A neuron responding to a short line can be considered to be binding the set of points that make up that line. The inputs and the behavior of such a neuron are probably initially determined by our genes (and by developmental processes) that have evolved out of the experiences of our distant ancestors. Other forms of binding, such as that required for the recognition of familiar objects, such as the letters of a well-known alphabet, may be acquired by frequently repeated experience—that is, by being overlearned. This probably implies that many of the neurons involved have as a result become strongly connected together.* These two forms of fairly permanent binding could produce neurons that, collectively, could respond to many objects (such as letters, numbers, and other familiar symbols), but there are not enough neurons in the brain to code for the almost infinite number of *conceivable* objects. The same is true of language. Each language has a large but limited number of words, but the number of possible well-formed sentences is almost infinite.

The binding we are especially concerned with is a third type, being neither determined during early development nor overlearned. It applies particularly to objects whose exact combination of features may be quite novel to us, such as seeing a new animal at the zoo. It is unlikely that all the actively involved neurons will be strongly connected together, at least in most cases. This binding must arise rapidly. By its very nature, this third type is largely transitory and must be able to bind visual features together into an almost infinite variety of possible combinations, although it may be able to do this for only a few combinations at any one time. If a particular stimulus is repeated frequently, this third type of transient binding may eventually build up the second, overlearned type of binding.

Unfortunately, we don't yet know how the brain expresses this third type of binding. What is especially unclear is whether, in focused awareness, we are conscious of only *one* object at a time, or whether our brains can deal with several objects simultaneously. We certainly

---

*Recall that most cortical neurons have many thousands of connections and that initially many of these may be weak. This would imply that one can only learn something easily and well if the brain is already roughly wired in the right way.

appear to be aware of more than one object at once, but could this be an illusion? Does the brain really deal with several objects one after another in such rapid succession that they appear to be simultaneous? Perhaps we can *attend* to only one object at a time but, having attended, can briefly "remember" several of them. Since we do not know for certain we have to consider all these possibilities. Let's assume first that the brain deals with one object at a time.

What sort of neural activity could correspond to binding? Of course the neural correlate of consciousness may involve only one particular type of neuron: for example, one sort of pyramidal cell in one particular cortical layer. The simplest idea would be that awareness occurs when some members of this special set of neurons fire at a very high rate (e.g., at about 400 or 500 Hertz), or that the firing is sustained for a reasonably long time. Thus "binding" would correspond to a relatively small fraction of cortical neurons, in several distinct cortical areas, firing very fast at the same time (or for a long period). This is likely to have two consequences: the rapidity or duration of firing would increase the effect produced on the neurons to which this active set of neurons projects, neurons corresponding to the *implications* (or the "meaning") of the object that one was aware of at that moment. In addition, the rapid (or sustained) firing might activate some form of very short term memory.

This simple idea will not work if the brain has to be aware of more than one object at precisely the same moment, and even with one object the brain may have to distinguish between figure and ground. To grasp this, suppose that in the visual field, near the center of vision, there is just a red circle and a blue square. Then some of the neurons corresponding to awareness would be firing fast (or for a sustained period), some signalling "red," some "blue," and others signalling "circle" and still others "square." How could the brain know which color to put with which shape? In other words, if awareness corresponded merely to rapid (or sustained) firing, the brain might easily confuse the attributes of different objects.

There are several ways to get around this difficulty. An object may only enter vivid awareness if the brain "attends" to it. Perhaps the attentional mechanism can strengthen the firing of the neurons responding to one of the objects while weakening the activity of the neurons responding to other objects. If this were true, the brain would be able to deal with one object after another, as the attentional mechanism jumped from one object to the next. After all, this is what we do when we move our eyes. We attend first to one part of the visual field, then to another part, and so on. The attentional mechanism we need would have to be faster

as we lie beneath the stars
we realize how small we are

than this and operate *between* eye movements, when the eyes are stationary, since we can see several objects without moving our eyes.

A second alternative is that, in some manner, the attentional mechanism makes different neurons fire in somewhat different ways. The key idea here is that of *correlated* firing.* It is based on the idea that what matters is not just the average rate of firing of a neuron but the *exact moments* at which each neuron fires. For simplicity, let us consider only two objects. The neurons associated with the properties of the first object will all fire at the same moment, in some sort of pattern. The neurons associated with the second object will also all fire together but at different moments from the first set.

An idealized example may make this clearer. Suppose the neurons of the first set fire very fast. Perhaps this set will also fire again, say 100 milliseconds later, and again 100 milliseconds after the second burst, and so on. Suppose the second set also produces a set of fast spikes every 100 milliseconds or so, *but at times when the first set is silent.* The other parts of the brain will not confuse the neurons in the first set with those in the second set, because the two sets never fire at the same moment[†] (see Fig. 57).

The basic idea here is that spikes arriving at a neuron at the same moment will produce a larger effect than the same number of spikes arriving at different times.[‡] The theoretical requirement is that the firing of the neurons in each set should be strongly correlated with each other, while at the same time firing of neurons in different sets should be weakly correlated, or not at all.**

*   *   *

*This was proposed by Christoph von der Malsburg in a rather obscure paper in 1981, although Peter Milner and others had mentioned it before then.

[†]Of course, the axonal spikes of one set need not be exactly synchronous with each other, since a neuron that receives the spikes will spread their effects somewhat over time as the changes of electrical potential travel down its dendrites to the cell body. In addition, there may be somewhat different time delays involved as the spikes travel along the many different axons. Thus the times of firing within a set may only have to be simultaneous within a few milliseconds or so.

[‡]A slightly more elaborate theory would arrange the inevitable time delays in axonal transmission so that synapses farther from the cell body would get their input a little sooner than those nearer to it. Then, because of the small differences in dendritic delays, the two signals will have their maximal effects on the cell body at the same time. Even more elaborate theories would involve the timing of inhibitory effects produced by the local inhibitory neurons. All of these qualitative considerations can be made more quantitative by careful simulations, in a computer, of the way an individual neuron might behave under these circumstances, allowing for the time delays and so on.

**The firing is unlikely to be as regular as that shown in Figure 57.

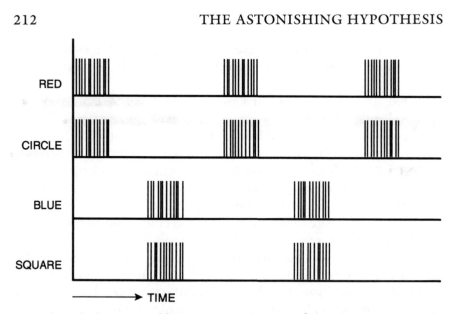

Fig. 57. Each short vertical line represents a neuron firing at some moment in time. The first horizontal line shows the firing of a neuron that signals "red." The next line, one that signals "a circle," and so on. The brain can deduce that the circle is red, not blue, because the red neuron fires at about the same times as the circle neuron, whereas the blue neuron fires at very different times. This is expressed by saying that the firing of the "red" and the "circle" neurons are *correlated* (as is that of the "blue" and the "square" neurons), whereas the *cross-correlation*, for example, between "red" and "square," is zero. (The example has been grossly oversimplified for didactic purposes.)

Let us now return to our main problem. This is to locate the "awareness" neurons and to discover what it is that makes their firing symbolize what we see. This is like trying to solve a murder mystery. We know something about the victim (the nature of awareness) and we know various miscellaneous facts that may be related to the crime. Which approaches look the most promising and how should we follow them up?

The most direct set of clues would be any evidence that caught the suspect in the act. Can we find neurons whose behavior always correlates with the relevant visual percept? One way to do this would be to set up situations (such as viewing the Necker cube, described in Chapter 3) in which the visual information coming into the eyes remains the same but the percept changes. Which neurons change their firing, or style of firing, when the percept changes and which do not? If a particular neuron does not follow the percept, this provides it with an alibi. On the other hand if its firing does correlate with the percept we still have to decide whether it is the actual murderer or only an accomplice.

Suppose we try another tack. Can we pin down the crime to a particular town, or a particular district or apartment building? This would make our search more efficient. In our problem, can we say roughly where in the brain the awareness neurons for vision are likely to be located? Obviously we suspect the neocortex, although we cannot entirely neglect its close neighbors, the thalamus and the claustrum, nor even that older visual system, the superior colliculus, to say nothing of the corpus striatum and the cerebellum. Visual awareness is unlikely to reside in areas such as the auditory cortex, so we can confine most of our attention to the many visual cortical areas shown in Figure 48. Perhaps we might find evidence that some areas are more heavily involved than others.

This will not pin down the murderer but it may lead us in the right direction. Is the criminal likely to be any particular type of person—a powerful man, for example, a disturbed teenager, or a gang? In our case, what kind of neurons are likely to be involved? Excitatory neurons? Inhibitory neurons? Stellate cells, or pyramidal cells? If they are in the cortex, in which layer or layers are they to be found?

Another tack would be to see if there were any forms of communication that might give the game away. If it is the work of a gang, did they use a cellular telephone in a car? In neural terms, does awareness depend on some particular form of neural circuitry that only occurs in special places in the brain?

Perhaps one should look for a motive for the crime. What benefit was the murder to the murderer? Did he profit financially and, if so, where was the money sent? If we could look there, we might be able to track backwards to the murderer. In neural terms, to what parts of the brain is the visual information dispatched? And how are these parts connected to the visual areas of the cortex?

Alternatively, one could ask if there is any special behavior that might lead us to the suspect. This might be correlated firing between groups of neurons, or perhaps rhythmical or patterned firing of one sort or another. If a gang is suspected, who is likely to be the leader, who decides what the gang should do? We believe that awareness often involves the brain making decisions as to which interpretation is the most plausible. There may be a winner-take-all mechanism involving certain sets of neurons. If we could spot such a mechanism the neuronal nature of the winners might point us toward the awareness neurons. Was some particular weapon used? As mentioned earlier, we strongly suspect that very short term memory may be an essential feature of awareness. Also that some form of attentional mechanism may help to produce vivid awareness, so anything we can learn about how they work neuronally might lead us in the right direction.

In short, there are a number of experimental approaches that might conceivably lead us to the neurons and to the behavior that we are looking for. At this stage we can't afford to neglect any lead that looks even remotely promising, since we have a difficult problem to solve. Let us now examine in more detail the nature of these different approaches.

15

Some Experiments

"Through purely logical thinking we can attain no knowledge whatsoever of the empirical world."
—Albert Einstein

A particular neuron in a monkey's brain may be sensitive to the color of a particular patch of the visual field. But how can we be sure that it is directly involved in its perception of color? It might, for example, be part of a system that merely draws the brain's attention to that part of the visual field. If so, a person who had lost, by brain damage, the true color neurons might see only in black and white, and yet his attention might be drawn to a patch of color.

This is not an abstract possibility. Alan Cowey and his colleagues at Oxford have studied in great detail[1] a person who, because of brain damage, has lost the percept of color (in ordinary language, he cannot see color but does see in black, white, and shades of gray). They have shown that the subject can say whether the colors of two small, colored squares (adjusted to be equally bright) are the same or different, provided the two squares touch each other. This in spite of the fact that such a person stoutly denies he can perceive their color. If the squares do not touch, he cannot do the job and his performance falls to chance levels. This shows rather clearly that some information about color is available to the brain without the person being aware of color.

✳   ✳   ✳

William Newsome, at Stanford University, has done a series of brilliant experiments attempting to find if the response of certain neurons is related to what a monkey sees. The cortical area chosen was MT (sometimes called "V5"), the one whose neurons respond well to movement and only indirectly, or not at all, to color (see Chapter 11). It had already been shown that damage to this area made it difficult for a monkey to respond to visual motion, although this handicap often wore off after a few weeks (the brain probably learns to use other pathways).

Following on earlier work by others, Newsome and his colleagues first studied how individual neurons in MT responded to specially chosen motion signals.[2] These signals consisted of a pattern of random, rapidly changing dots on a TV screen. In one extreme condition all the transient dots moved in one direction. This motion could easily be seen. At the other extreme their average motion was zero, as occurs sometimes in the "snow" seen on TV sets tuned between stations. The observer had to signal whether the motion was in the chosen direction or its opposite. When the average motion was zero, his performance was at chance.

Newsome and his colleagues used various mixtures of these twinkling patterns. If all the motion was in one direction, the monkey (or a man) could always signal correctly what that direction was. With only some of the dots moving in one direction and the others moving randomly, the observer sometimes made mistakes. The fewer the proportion of directional dots, the more mistakes were made. By trying various proportions, it was possible to plot a curve showing how the observer's accuracy varied with the percentage of directional dots.* For each neuron whose electrical activity they studied, they pretended, using a special mathematical method, that it was that neuron that was making the decision about direction in the most efficient way.

In all, they studied over two hundred different neurons. About a third of them did as well as the monkey. Some did rather worse, but others discriminated motion rather better than the monkey did. Why, then, did the monkey not respond more successfully, since its brain had neurons in cortical area MT that could do the job better? The most likely answer is that the monkey cannot choose just one neuron (the most discriminating one) to control its response. Its brain must use a population of neurons. Exactly how it does this is not yet understood.

*Such a curve is called a "psychometric curve."

What the experiment does show is that the visual information needed to make the choice is present in the behavior of MT neurons, so that one cannot say that those neurons could not do the job. Unfortunately, this does not prove that they actually do it.

Newsome's next experiment takes us one step further.[3] He and his colleagues asked the question: If we stimulated appropriate MT neurons to make them fire when making a difficult discrimination, will the monkey's behavior improve?

It is not technically easy to stimulate just one neuron. Fortunately the neurons in cortical area MT that respond in similar ways—that is, to one particular direction of motion in one particular part of the visual field—tend to be clustered together. By electrically stimulating a small patch of cortex near the neuron being studied, there is a good chance that all these somewhat similar neurons will be stimulated together.

In all they did sixty-two experiments. In about half of these the monkey's motion discrimination was appreciably improved by the stimulating current. This is a really striking result. It means that we can improve the way the monkey responds to a particular visual stimulus by exciting neurons in the appropriate place in the visual cortex. It had to be at that particular place because if the stimulating current was applied at other places in cortical area MT it had little or no effect on the monkey's performance of this particular task.

Does this imply a small patch of MT contained the neural correlate of *seeing* that kind of motion? This is certainly plausible but there are several difficulties about drawing this conclusion with certainty.

One possible objection is that the monkey, while showing the appropriate behavior, was not really seeing anything. It was just responding like an automaton, without visual awareness. The only way to answer this objection conclusively is to fully understand the visual system in both monkeys and humans; so, for the time being, we have to assume that a monkey has visual awareness until the evidence suggests otherwise.

It could be argued that while a monkey does have visual awareness it does not have it for this particular task. This is unlikely because for this task the choices of a monkey and a man are very similar—that is, the psychometric curves are much the same. It is not the case that the monkey's performance is far inferior to the man's. Thus it is plausible that their brains are using similar mechanisms; yet there is one further difficulty.

If a person performs such a task repeatedly, his behavior often becomes almost automatic. He may report that he barely glimpsed the movement, if at all, but in spite of this his choice may be well above chance. It is more difficult to train a monkey than a man, as you cannot describe the task to him verbally. Newsome's monkeys were probably overtrained, and for this reason their behavior may have become somewhat automatic, with little if any contribution from visual awareness.

I doubt myself whether this objection is really a serious one, since if most of the twinkling dots move in the same direction the movement is clearly seen by us, and almost certainly by the monkey as well. Unfortunately, the stimulating current usually makes little difference in such cases, since the performance is already almost perfect. It may be possible to train the monkey to discriminate the direction of movement using a different moving stimulus, such as an oriented bar, and then test him on the moving dot task before he becomes overtrained. Such an experiment is not easy, as there are a number of hazards, but it might be worth a try.

A more serious objection is that although the behavior of the neurons in cortical area MT appears to be correlated with the monkey's discrimination, and therefore probably with its visual awareness, it does not follow that these particular neurons are the real seat of awareness. They may, by their firing, influence other neurons, perhaps elsewhere in the visual hierarchy, that are the true correlates of awareness.

The only answer to this is to study other cortical areas. If we cannot find neurons with similar powers of discrimination elsewhere, then the case for the MT neurons would be strengthened. In the long run we cannot hope to pin down visual awareness until we understand much more about all the visual areas and, in particular, how they interact together. Nevertheless, Newsome's experiments are a very important first step in that direction.

If a neuron fires to some stimulus in the visual field, we naturally suspect that it may be the neural correlate of our percept of that stimulus, although, as just explained, this certainly does not follow automatically. Are there more powerful methods that might narrow down the search for the awareness neurons? Can one find situations in which the visual input is constant but the percept is changing? We could then try to find which neurons in the monkey's brain followed the input and, more to the point, which followed the changing percept.

One obvious case is the Necker cube (see Fig. 4). Here the figure stays the same, but we see it in 3D, first in one way, then in the other, and so on. At this point in time, it is not obvious where to look in the brain for the percept of a 3D cube. We need to study something that is easier to locate in the monkey's visual system.

One very attractive possibility is based on a phenomenon known as binocular rivalry. This occurs when each eye has a different visual input relating to the same part of the visual field. The early visual system on the left side of the head receives an input from both eyes but sees only that part of the visual field to the right of the fixation point. (Conversely for the right side of the head.) Two conflicting inputs are called "rivalrous" if you do not see the two inputs superimposed but first one input, then the other, and so on in alternation.

A very dramatic example of binocular rivalry, prepared by Sally Duensing and Bob Miller, can be seen at the Exploratorium at San Francisco.[4] In the demonstration at the Exploratorium the viewer puts his head in a fixed place and is told to keep his gaze fixed. By means of a suitably placed mirror (see Fig. 58) one of his eyes can look at another person's face, in front of him, while the other sees a blank white screen to the side. If the viewer waves his hand in front of this plain screen, at the same point in his vision where the face is, the face is wiped out! The movement of the hand, being visually very salient, has in some sense captured the brain's attention. Without attention the face cannot be seen. If the viewer moves his eyes, the face comes back again.

In some cases only part of the face disappears (Fig. 59). Sometimes, for example, one eye, or both eyes, will remain. If the viewer looks at the smile on the person's face, the face may disappear leaving only the smile. For this reason the effect has been called the "Cheshire Cat effect," after the cat in *Alice in Wonderland*. You can try it for yourself, using a simple pocket mirror. The results are really striking. The experiment works best if there is a uniform white background behind both the person being observed and the observer's hand.

So far this has not been tested on a monkey, but a much simpler experiment has been done at the Massachusetts Institute of Technology. Nikos Logothetis and Jeffrey Schall trained a macaque monkey to signal whether it is seeing the upward or the downward movement of a horizontal grating.[5] To produce rivalry, upward movement is projected into one of the monkey's eyes and downward movement into the other one in such a way that the two images overlap in the monkey's visual field. The monkey signals that he sees up and down movements alternatively, just as we would. Notice that the

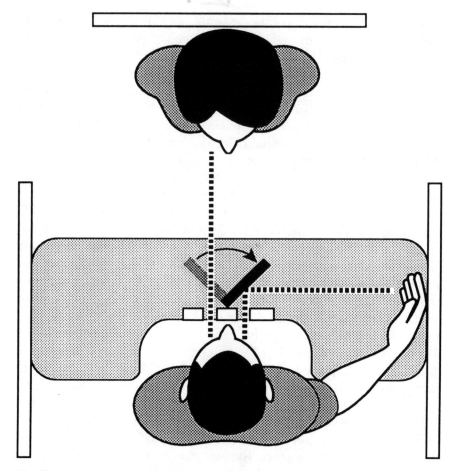

Fig. 58. A sketch of the Cheshire Cat exhibit used by the Exploratorium visitors. A double-sided mirror is provided so that the observer can conveniently use either eye for the erasing field. This arrangement allows the observer to use his own hands for erasing.

motion stimulus coming into the monkey's eyes is always the same, yet the monkey's *percept* changes every second or so.*

Cortical area MT is an area mainly concerned with movement but largely indifferent to color. What do the neurons in MT do in a short time when the monkey's percept is sometimes up and on other occasions down? The answer is that the firing of some of the neurons correlates

*The time intervals follow a gamma-distribution.

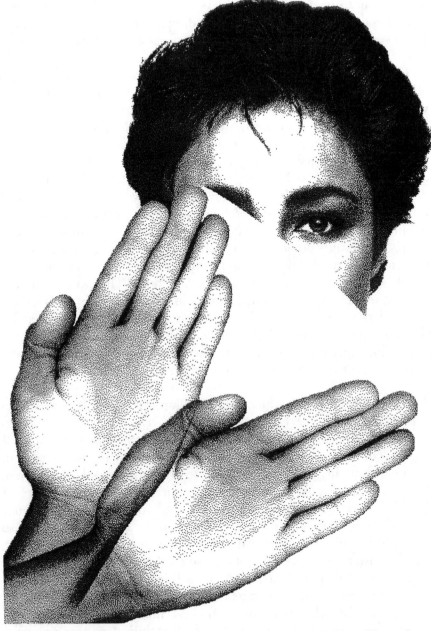

Fig. 59. The observer sits in the apparatus shown in Figure 58 and keeps his eyes fixed. If he waves his right hand so that its image in the mirror passes over the image of part of the other's face, that part of the face disappears. If he moves his eyes, the face appears again. (Note that the mirror turns his right hand into a left hand.)

with the percept, but for others the average firing rate is relatively unchanged and independent of which direction of movement the monkey is seeing at that moment. (The actual data are rather more messy than this simple description.)

This result makes it unlikely that *all* the neurons firing in the visual cortex, at one moment, are correlates of our visual percept, although it would be nice to have a few more examples of this. Unfortunately it does not pin down the awareness neurons precisely. As explained for Newsome's results, the real correlate might be the firing of neurons elsewhere in the visual hierarchy, influenced, at least in part, by the firing of those in MT. Ramachandran has suggested[6] that this rivalry may not be true movement rivalry, but really form rivalry, and that its real seat might be lower in the visual hierarchy, perhaps in cortical area V1 or V2. Again, even if some of the awareness neurons are indeed located in MT, the present results do not show exactly which neurons they are. Which cortical layers are they in? Which type of neuron are the ones that tend to follow the percept rather than the visual input? The same possibility that the monkey was overtrained, as was discussed for Newsome's results, is also a worry here, although this is unlikely because rivalry shows very little variation with training. But again, with all these reservations, these are important experiments. Pursued further they should lead us toward the explanation of visual awareness in neural terms.

Are there other cases in which the visual input is constant but the percept changes for one reason or another? Of course there are cases where the observer suddenly "sees" an object that his brain had not spotted before, as in the Dalmatian dog hidden in the picture in Figure 9, but this would not be easy to study in a monkey. A human being can say, "Look, I now see a dog, but I didn't see it before." It would be more difficult for a monkey to tell you this. Moreover, once seen, the dog is usually spotted straight away on a subsequent test, so it would be difficult to repeat the same experiment several times, which is usually necessary to obtain a scientifically secure result.

One possibility is to study the effects produced in the brain by images that fade from awareness because they have been stabilized on the retina. (Recall that normally the eye makes various tiny movements that prevent such fading.) Retinal stabilization was originally done by fitting on the eyeball a small, rather uncomfortable gadget that projected chosen patterns of light onto the retina. When the eye moved the pattern stayed in the same place on the retina, in spite of the eye movement, and so faded. A number of such experiments were done in the fifties, but few seem to have been done since, although a more elaborate (and more comfortable) device is now available to produce stabilized images.[7]

It might be thought that the fading process takes place mainly in the retina, and so would be of little interest to us, but this seems unlikely. These early studies showed that a complex pattern does not always fade as a whole.[8] A straight line does tend to act as a unit, but the several lines of a square or a triangle may fade independently. Jagged figures are less stable than rounded ones. What the Gestalt psychologists would have called a good figure is more likely to act as a unit than a poor figure. If the pattern is a capital B, with a crude squiggle scrawled over it, the squiggle is likely to fade before the B fades. This suggests that the fading is mainly taking place in the brain, not in the eye. It might be worthwhile trying to stabilize various patterns on the retina of an alert macaque monkey, having trained the animal to signal something of what he sees, and observe which neurons are influenced when that part of the image fades from awareness.

Another possibility is to study the dramatic experiment reported by Ramachandran and illustrated in Figure 19. This involved the apparent movement of two stationary misaligned, parallel lines, touching the locally blind region produced in humans by a small area of brain damage (called a "scotoma") in cortical area V1. It is possible that this could be studied in a monkey, if the monkey could be trained to signal moving versus stationary, aligned versus misaligned, and interrupted versus continuous. As far as I know, no one has yet attempted this.

A simpler experiment has already been done on the real blind spot in a monkey. (For the description of the psychology of our blind spot, see Chapter 3.) There is a region in the first visual area, corresponding to the blind spot, in which the cortex receives a direct input from only one eye, since the other eye has no photoreceptors to cover that area of the visual field. (Recall: Most of V1, on one side of the head, receives input from both eyes, although it deals with only the opposite [contralateral] half of the visual field.) It might be thought that the neurons in the blind spot region responded only to the signals from one eye. Surprisingly, this is not the case. Ricardo Gattass and his colleagues from the Universidade Federal in Rio de Janeiro have shown that, in the macaque monkey, some of them do respond to input from both eyes.[9] This unexpected input from the locally "blind" eye probably comes, directly or indirectly, from adjacent cortical tissue that receives inputs from both eyes. Wherever it comes from, they have shown that the neurons in the blind spot region of V1 respond by firing to signals, of the type described in Chapter 3, that produce filling-in. This, incidentally, decisively disposes of Dennett's argument (outlined in Chapter 4). This is a neat example of the general principle ⸜that if you vividly see some aspect of the visual scene, *there must be neurons firing whose activity explicitly symbolizes that feature.* (Another example of this principle is the neural response to subjective contours described in Chapter 11.)

This particular blind spot phenomenon tells us little more about the location of the awareness neurons than an ordinary example of neurons firing to a visual input. If it could be extended, as suggested earlier, to cases in which the percept changes to a constant visual input (as in Fig. 19), it could help us in our search.

Another approach is to study cases in which several distinct visual inputs produce the same percept, or at least certain elements of that percept. An example of this would be the experiments by Tom Albright and his co-workers at the Salk Institute on cortical area MT of the macaque (see also pages 151–153). They have shown that, even if the moving objects studied are fairly different, the responses to movement of some MT neurons have much the same character. For example, the movement of a patch of ripples across the visual field makes some MT neurons respond in much the same way as they would to a rigid bar moving in the same place and in the same direction. Although the patterns are different, their movement is much the same. (They call this "form-cue invariance.")

So far they have not shown whether there is anything special about the type, location, or firing behavior of such neurons. If these are awareness neurons, we might hope that, whatever the incoming signals, their firing (or some aspect of it) would always be correlated with the visual percept.

The evidence so far is so weak that it is reasonable to ask: Can one study exactly the same neuron when the animal is alert and then again when it is unconscious? It is difficult, for technical reasons, to do this if the animal is made unconscious by an anesthetic, but it has been done by comparing an alert cat with the same animal in slow-wave sleep.*

---

*In REM sleep, the brain waves closely resemble those of the awake brain, suggesting that, in REM, the brain is at least partly conscious, as we seem to be in our dreams. The brain waves in the slow-wave (non-REM) phase of sleep are quite different from those of the alert brain, and few, if any, dreams occur at this time. It is reasonable to assume, therefore, that we are usually unconscious during slow-wave sleep.

The neuroscientists Margaret Livingstone and David Hubel published such an experiment[10] in 1981. Most of the neurons they studied* were in cortical area V1. The animal's eyes were open, so that even in slow-wave sleep the neurons in V1 responded to visual signals produced by a computer on a screen in front of it. When they had recorded the responses of one particular neuron they woke the animal and tested it again with exactly the same stimuli it had viewed before.

When the animal was awake each neuron they studied responded in broadly the same way as when it was asleep—that is, if it preferred a certain oriented line in a certain position in the visual field, its preference was much the same whether it was awake or asleep, although the signal-to-noise ratio was usually better when it was awake.[†] However, quite a number of cells gave an increased rate of firing when the animal was awake, compared with that when it was asleep. This is perhaps not too surprising, but the interesting result was that there were more striking changes in response in the lower layers of the cortex (layers 5 and 6) than in the upper layers.

They confirmed this general result by using a chemical (radioactive 2-deoxyglucose) that showed the average activity (over a period of about half an hour) produced by visual stimuli in each of these layers, in one case when it was awake and in a comparison case, with a different radioactive isotope, when it was asleep. The result was broadly the same. The activity was markedly more in the lower cortical layers when the animal was conscious, while there was very little change in the upper layers.

This tempts one to a broad generalization that goes far beyond the present evidence. This is that the activities in the upper cortical layers are largely unconscious, while at least some of those in the lower layer correspond to consciousness. I must confess to being unreasonably fond of this hypothesis—it would be so pretty if it were really true—but as yet I cannot bring myself to embrace it wholeheartedly. There could be other reasons why the lower layers are less active in slow-wave sleep.

Can we learn anything about awareness by studying the mechanism of attention? Experimental work on the neural basis of attention has been going on for some time. Some of this has been done on alert monkeys by recording the firing of neurons in various parts of its brain

---

*They also looked at some neurons in the LGN.

†That is, the firing rate to the stimulus was higher compared to the background rate of firing of the neuron.

when the monkey is performing certain visual tasks. Experiments have also been done using PET scans on humans, as mentioned in Chapter 8. I shall not attempt to recount all these experiments; instead, I shall briefly describe only one of them and the sort of results it has given.

Robert Desimone and his colleagues at the National Institute of Mental Health in Bethesda, Maryland, have trained a monkey to fixate a point to one side of a visual display and to pay attention (without moving its eyes) to one feature of the display or another.[11, 12] Various signals are then flashed and the experimenter studies the firing of a particular neuron in cortical area V4 that responds to a display in that position. Neurons in V4 are interested in color. Suppose the neuron being studied responded to a red bar (in a certain orientation) but not to a green bar. (Of course, other neurons in V4, not being studied at that moment, will respond to a green bar but not to a red one.) Each time the display contains two colored bars, one red (the effective stimulus for that neuron) and one green (the ineffective stimulus), both of them inside the neuron's receptive field. When the monkey attended to a position occupied by the red bar the neuron fired as well or better* than when the monkey was not attending. However, in those trials during which it attended to the green bar, the *red-sensitive neuron fired less*. Attention, therefore, is not just a psychological concept. *Its effects can be seen at the neuronal level*. The monkey's attention could make neurons that are sensitive to the attended stimulus fire more when its attention was on it and fire less when its attention was elsewhere, even though the position of the eyes and the incoming visual information *was exactly the same in both instances*.

They described their results this way:

> Neurons in area V4 . . . have receptive fields so large that many stimuli typically fall within them. One might expect that the responses of such cells would reflect the properties of *all* stimuli inside their receptive fields. However it has been found that when a monkey restricts its attention to *one* location within a V4 . . . cell's receptive field, the response of the cell is determined primarily by the stimulus at the attended location, *almost as if the receptive field "shrinks" around the attended stimulus*. [Italics added.]

I will not attempt to describe their more detailed results, since they are not easy to follow. They show that a simple theory of an attentional

---

*If the task was easy, the firing was much the same. If the color discrimination was made more difficult, then attention increased the firing rate.

spotlight is unlikely to be correct. What more complicated mechanism is needed to explain them has not yet been established.

Is the thalamus involved in attention? The thalamus ("the gateway to the cortex") has many fairly distinct regions, some of which are interested in vision. The main pathway from the eyes to the cortex is through the LGN (the Lateral Geniculate Nucleus)—a part of the thalamus (described in Chapter 10). The other visual thalamic areas are in a region of the thalamus called (in primates) the "pulvinar,"* which is a large thalamic nucleus, considerably bigger than the LGN.

David Lee Robinson and his colleagues at the National Eye Institute in Bethesda have done a number of experiments on parts of the monkey's pulvinar. The features to which the (retinotopic) pulvinar neurons respond appear to depend upon their inputs from the visual cortex and not on those from the superior colliculus.[†]

If, by chemical means, inhibition is increased in one small region of the pulvinar, the monkey has more difficulty in shifting attention, whereas reducing inhibition makes it easier to do so. Experiments by others show that the pulvinar acts to suppress inputs for irrelevant events. Studies on three human patients with thalamic damage suggested that they had difficulty in engaging attention. PET scans on normal humans have shown that there was increased activity in the pulvinar when there were distractors in a visual task. These distractors made the subject use more attention to do the job. All these results (for a review see reference 13) strongly suggest that these parts of the thalamus are intimately involved in various aspects of visual attention.[‡]

This is clearly a rich field for further work. The exact connections of each pulvinar area (glossed over above) need further detailed study—for example, in what way do the several retinotopic areas differ in their

---

*The pulvinar has three main parts and one smaller one. Two of these (the inferior and the lateral regions) are retinotopic—each has one or more maps of the visual field. They have connections, in both directions, to most of the early visual areas and also receive strong, nonreciprocal connections from the superior colliculus. The third one, called the medial pulvinar, does not have a retinotopic map and is connected (in both directions) mainly to the parietal and frontal cortex. It probably responds somewhat to the other senses, not just to vision, and may be more cognitive and less involved with vivid visual awareness.

†Recall that the superior colliculus is closely involved with the control of eye movements, another form of visual attention. The collicular inputs to the pulvinar, on the other hand, seem more concerned with the salience of the different parts of the visual field.

‡This has also been proposed by Anderson and Van Essen[14] as part of their shifter-circuit theory.

connections? Can we say with more precision just how each particular part of the pulvinar influences attention and how it interacts with the neurons in the various cortical areas to which it is related? Further experimental work should answer these questions. (I discuss some speculative ideas about the different regions of the pulvinar in Chapter 17.)

How much shall we learn about the neural basis of visual awareness by studying the thalamus? Since attention is important for awareness we would be foolish to neglect it. To discover how we see things we must not only understand the workings of the neocortex but those of the LGN and the pulvinar as well.

Are there any relevant experiments that could be done on a man rather than a monkey? This would have the advantage that the subject could report verbally what he was experiencing, which a monkey cannot do. It is rarely possible, for ethical reasons, to stick electrodes into a person's brain, but this sometimes has to be done for medical reasons. It is also possible to study brain waves, from the outside of the head, but these are usually rather more difficult to interpret.

Such an approach has been pioneered by Benjamin Libet, working at the University of California, San Francisco. He prefers to work on humans because he is reasonably certain that other human beings are conscious. (He feels less confident in the case of a monkey.) Any experimental work on consciousness has, in the past, been looked on with grave suspicion, not only by psychologists and neuroscientists but also by the medical profession. Almost the only interest surgeons and anesthetists have had in the subject has been how to anesthetize a patient during an operation in such a way that he is unaware of what is going on, partly to spare the patient pain and partly to prevent him from suing them. (Libet told me that he wisely did not embark on his experiments on consciousness in alert people until he had obtained the security of academic tenure.)

Libet's main work has been concerned with certain brain waves that precede a voluntary movement and how these events in the brain relate to the apparent time of the subject's awareness of intending or wanting to move.* His results suggest that for this type of conscious awareness there must have been brain activity for a certain minimum

---

*I shall not describe these experiments for two reasons: They do not relate directly to the visual system, which is my main concern, and they are difficult to interpret and have led to controversy, so that if they are to be discussed at all they have to be described at some length and this would be too much of a distraction. They are more relevant to the question of Free Will, discussed briefly in the Postscript.

time, perhaps for a hundred milliseconds or so. The exact value probably depends upon the strength of the signal and the circumstances.

His other work, which is more recent, concerns the effects of stimulating part of the thalamus (the ventrobasal complex) concerned with
sensations such as touch and pain. This was done to people who have
had electrodes implanted there for the relief of intractable pain. I shall
describe these experiments[15] because even though they are not concerned with vision they may be relevant for the interpretation of blindsight (as discussed in Chapter 12).

The subject's thalamus was given a certain amount of stimulation.
He (or she) then had to decide (guessing if necessary) when the stimulus occurred. More precisely, did it occur when a particular light was
on for one second, or when a different light was on in the subsequent
second. The subject indicated his choice by pressing one of two buttons provided. If his brain had no knowledge of when the stimulation
occurred, the subject would have to guess and so would be correct, on
average, 50 percent of the time. After the stimulus and his response
were over, he had to press one of three other buttons to indicate
whether he had been aware of the stimulus. The subject was to press
button one if he had felt the sensation (in the usual place), even if it
was very brief. If he was uncertain, or thought that he might have
experienced something, he was to press button two. If he simply felt
nothing at all, he was to press button three.

The experiments conducted by Libet and his colleagues were complicated, so I shall only describe the broad result. The stimulus consisted of electrical pulses at the rate of 72 per second; on different
trials different numbers of pulses were delivered, always of the same
amplitude. The results showed that even when the train of pulses was
too brief to elicit awareness, the subjects could perform above chance.
To become aware of the stimulus (even if this awareness was somewhat uncertain) required a significantly longer train.

Libet and his colleagues interpreted this as implying that a certain
duration of the pulse train was needed for awareness. Unfortunately,
they did not, in these experiments, systematically vary the intensity of
the stimulus, but this and earlier work had suggested that an increase
of intensity of a train of a fixed duration can change the subject's
response from nonawareness to awareness. In short, in the somatosensory system, a weak or brief signal can influence behavior without producing awareness, while a stronger or longer one of the same type can
make awareness occur. The exact neural behavior produced by
stronger or longer stimuli remains to be determined.

This result means that, when trying to explain blindsight, we cannot ignore a similar type of explanation—that is, that the pathway from the LGN to areas such as V4 may be too weak to produce visual awareness yet still be strong enough to have some influence on the person's behavior.

Although the experiments described in this chapter have not yet led to any strong conclusions about the precise neural correlates of visual awareness, they do show that an experimental approach to one aspect of consciousness is possible. If pursued vigorously such experiments might eventually lead to a solution to the problem.

Another, parallel, approach is to try to guess the general nature of the answer, if only as a guide to further experiments that otherwise might not be performed. Some of these speculative ideas are outlined in the next chapter. They do not (as yet) form a single coherent set of ideas, but rather a hodgepodge of rather tentative suggestions, although some of them can be fitted together in a plausible way, as we shall see.

16

Mainly Speculation

"Give me a fruitful error any time, full of seeds, bursting with its own corrections. You can keep your sterile truths for yourself."

—Vilfredo Pareto

The experiments sketched so far should help to identify the neurons whose firing, at a given moment, correlates with some aspect of the visual percept. There are about half a billion neurons in the visual cortical areas on one side of a monkey's head. Have we any clues to guide us toward the neurons we are looking for?

One possibility is that all these neurons are potentially capable of being awareness neurons, although only a small subset of them may be fulfilling this role at any particular moment. This seems unlikely because of the behavior of neurons during binocular rivalry (discussed in the last chapter), yet there is always the possibility that those neurons that were not awareness neurons then may play that role on other occasions. Another more probable scenario is that there may be several rather different forms of visual awareness, perhaps a very transient one for simple features, a somewhat more persistent one for vivid visual awareness, and perhaps a further form that is indeed visual but does not correspond to the visual "picture" we appear to have in our heads. I have already touched on this when outlining David Marr's ideas in Chapter 6 and Jackendoff's ideas in Chapter 14. To make things simpler, let's concentrate, for the moment, on the vivid visual awareness that Jackendoff equates loosely with Marr's 2½D sketch.

One striking feature of our internal picture of the visual world is how well organized it is. Psychologists may delight in showing us that it is not quite as "metrical" as we often think it is—that is, our judgment of relative sizes and distances is not always as accurate as an engineer's drawing would be, but we seldom get things jumbled in space when seeing them under ordinary conditions. It is true that the real external world is constantly there, so the brain can refer to it to check any provisional judgments it may have made, but even so, when our brain makes a symbolic representation of the visual world in front of us, that representation is very well organized spatially.

This would not be especially surprising if the neurons at all levels of the visual hierarchy were very fussy about exactly where the feature they respond to is located in the visual field, but as we have seen, this is not the case. A neuron that fires especially well to a complex object, such as a face, responds almost equally well whether the face is directly in the center of the animal's gaze or somewhat to one side, or even tilted a little from the normal upright position. This makes good sense. There could hardly be a separate neuron for every high-level feature in every possible position. There would not be enough neurons to go around.

On the other hand, the neurons in the first visual area, V1, are indeed fussy about the exact position of the relevant feature (such as orientation, movement, color, disparity, etc.) in the visual field. They can afford to do this because these features are relatively simple and stereotyped and because V1 is a large cortical area. It also helps that so many of the neurons in V1 are concerned with features near the center of gaze.

In 1974, the psychologist Peter Milner published a very perceptive paper* in which he argued that, because of the reasons given above, early cortical areas (such as V1) would have to be intimately involved in visual awareness as well as the higher cortical areas. He suggested that this could be implemented by a mechanism involving the numerous backprojections from neurons higher in the visual hierarchy to those lower down.[1]

---

*This paper, which he wrote during a sabbatical, had largely dropped out of sight. Neither Christof Koch nor I had ever heard of it. Fortunately, in 1991 we were at a meeting in Arizona at which Peter was present, and he told us of the existence of this almost forgotten article. In this paper he also proposed the idea of correlated firing to solve the binding problem. Over the years people have suggested somewhat similar functions for these back pathways, including Stephen Grossberg, Antonio Damasio, Shimon Ullman, and others.

The exact functions of these backprojections are still not known. As they are connections between cortical areas they all come from neurons that transmit excitation. A key question is: How strong are they? Here opinions differ, but it is possible that they are not normally strong enough to make neurons fire rapidly, although they are strong enough to modulate any firing produced by other inputs. This might imply that their effects are too weak to influence several distinct stages in succession. If area C projects back to area B, and area B, in turn, projects back to area A, one might wonder whether what happens in C would have much influence on A, unless there was a direct pathway back from C to A, rather than an indirect one through B. Diagrammatically we have

(showing only the back pathways). Can C influence A? Or do we need the extra pathway (shown above the other two) to do this?

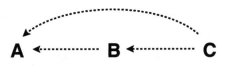

So we could ask: Which cortical areas (in the monkey) project directly back to V1?

Consulting the connection diagram (Fig. 52) we see that (almost) all the visual areas up to the level containing V4 and MT do indeed send back a direct connection to V1, whereas most of the upper ones in the hierarchy do not. Does this imply that only the neurons in the lower part of Figure 52 are directly involved in vivid visual awareness?

Since cortical area V2 is also large and also fairly retinotopic, an alternative might be that we need consider only those areas that project back to V1 or V2. This brings in further cortical areas but not the inferotemporal areas (those with IT in their initials).

I believe these ideas may have some truth in them, but the argument is too flimsy to be relied on as a certain guide as to where to look. It hints but it does not convince. Moreover, very recent work[2] suggests that more cortical areas project back to V1 than was originally thought. In such circumstances the best thing is to keep it in mind as the subject develops but not to put too much faith in it. At this stage, what is important is to find out more about the anatomy and the behavior of the many backprojections in the cortex.

Another possible tack is to ask whether awareness involves, in some sense, the brain talking to itself. In neural terms this might imply that reentrant pathways—that is, one that, after one or more steps, arrives back at the starting point—are essential, as Gerald Edelman has suggested.[3] The problem here is that it is difficult to find a pathway that is not reentrant. We have seen that this criterion might suggest that the hippocampus is the true seat of consciousness (it is reentrant since it gets most of its input from the entorhinal cortex and sends most of its output back there) and yet this is not the case. This negative result means that we must use the reentry criterion with care.

The simplest form of reentrant pathway would be between just two cortical areas. For example, if area A projected to area B, and also B projected to A; but this almost always happens, so this will not help us much. Can we make the idea of reentry more precise and therefore more useful?

Recall that, for many cortical areas, if A projects into layer 4 of B, then B does not project into layer 4 of A. The backprojection avoids that layer. We could symbolize this as

$$A \xleftarrow{\hspace{1.5cm}} \underset{\longrightarrow}{} B$$

where the solid arrow means "into layer 4." This suggest that we should look for the far fewer cases where two cortical areas do project reciprocally into layer 4—that is, using the above convention,

These are found (but not always) between cortical areas at the same level in the hierarchy of Figure 52. An obvious example is MT, V4, and V4t.

This idea looks quite attractive to me. It would be easy to produce theoretical arguments that could give it an air of intellectual respectability. Unfortunately the exact neural details of these alleged reciprocal layer 4 connections have not yet been studied carefully enough. It is certainly an idea to keep an eye on.

Let us try a rather different approach. So far I have talked mainly about cortical areas. Can one go a step further and try to guess which cortical layers might be involved in symbolizing awareness, or even which type of neuron, in one or more of those layers, might be involved? Here we do have a few crumbs of evidence.

There is one type of cortical neuron that attracts one's attention. These are some of the pyramidal cells in layer 5. They are the only neurons that project right out of the cortical system. (By the cortical system I mean the cerebral cortex and such regions as the thalamus and the claustrum that are very closely associated with it.) It could be argued that what one should send to other parts of the brain are the results of the neural computations. I have argued that it is plausible that visual awareness corresponds to a subset of these results. This makes one wonder about these particular pyramidal cells. Do they have any other unusual properties? (It is, in the long run, the agreement between many apparently different aspects of an object or concept that produces what scientists call "proof.") In fact, some of these neurons can fire in a special way. A number of neuroscientists have found that such neurons* tend to be "bursty." They injected current into many different individual neurons in cortical slices and found that their pattern of firing falls into three classes.[4] The first corresponds to that of inhibitory neurons. The second to most pyramidal cells, but the third, the neurons of which tend to fire in bursts under these circumstances, appear to be mostly the larger pyramidal neurons in layer 5. The apical dendrites of these large neurons extend to the top layer of the cortex (layer 1), where they probably receive inputs (among others) from the backprojections mentioned earlier.

All this rather sketchy evidence makes one wonder whether these particular layer 5 pyramidal cells are closely involved in awareness. Yet even if layer 5 pyramidals do express the "results" of the cortical computations, it does not follow that all of them, in all areas, produce some form of awareness when they fire. Something more may be needed to do this; for example, some special form of short-term memory, such as the reverberatory circuits discussed later in this chapter.

These ideas, while speculative, do underline the importance of knowing which layer and, if possible, which type of neuron a neuroscientist is recording from when he reports some experimental results. This is often technically difficult when alert animals are being studied, although new, more elaborate, methods may make this easier.

---

*These neurons do not produce axonal spikes in a completely regular way, nor at random time intervals; instead, they tend to produce short bursts of several spikes at a time, with longer intervals having only a few or no spikes between the bursts.

There is a more general argument that suggests we should pay close attention to cortical layers. Although the dendrites and axon of a neuron often extend over several layers, the layer in which its soma (the cell body) is located is probably determined genetically during normal embryonic development. (The *details* of the neuron's connections, on the other hand, are greatly influenced by its experience.) If there are indeed special types of cortical neurons whose firing correlates with what we see, then we might expect the soma of such neurons to be located in only one or a few cortical layers or sublayers.

The content of visual awareness is the result of the brain's attempt to make sense of the information coming into the eyes and to express it in a compact and well-organized manner. There would be little point in doing this unless it were of some real use to the organism. It is likely to be needed in several fairly distinct places. To which parts of the brain should this information be sent? Two obvious places are the hippocampal system (involved in the temporary storage, or coding, of episodic memory) and the motor system, especially its higher, planning levels. Can we determine the locations of visual awareness in the cortex by tracking the connections backwards from these two destinations?

Unfortunately, at the moment this approach raises more difficulties than it solves. Visual awareness is likely to be combined at some stage with information from the other senses, such as audition and touch. When you drink a cup of coffee you are aware of both the look and the feel of the cup, and the smell and taste of the coffee. The higher visual areas do indeed project into areas of the cortex that are polysensory. What is not so clear is whether the type of visual awareness that is sent to the hippocampus and the motor system is related more to the vivid surface awareness of the 2½D sketch or the less pictorial information in the 3D model. Perhaps both are needed.

The anatomical connections between the visual and polysensory regions of the cortex and the hippocampal formation are now fairly well known (see Fig. 52). They show clearly that visual areas such as V4 and MT do not project there directly, nor do the inferotemporal regions. The visual information has to go through other cortical regions to get to the hippocampus. Unfortunately, our present knowledge of the neural responses in these regions is only rather sketchy and needs further study.

The pathways to the motor areas of the cortex have been studied to some extent, but much further work is needed on them. Moreover, there are other, more indirect routes to the motor cortex. There is a massive pathway from all over the cortex to the corpus striatum.

(Interestingly, this comes from some of the pyramidal cells in layer 5.) From there the information goes to parts of the thalamus and from there to the various motor or premotor areas of the cortex. There is also a pathway from the cortex to the cerebellum and back again to the thalamus and so to the cortex. Some of these pathways may be involved in "unconscious," rather automatic actions. Many more experiments will be needed on these parts of the brain if we are to understand the various types of visual (and other) awareness.

One characteristic of an awareness neuron is that its firing is often likely to be the result of a decision by the neural networks involved. Making an equitable compromise can be a linear process but making a sharp decision is a highly nonlinear one. Electing the president of the United States is a nonlinear process, whereas election by proportional representation is more nearly a linear one, at least after each individual vote has been cast. Neurons and, by extension, neural networks can behave in highly nonlinear ways, so there is no difficulty here in principle.

For neurons, the mechanism is likely to be a winner-take-all process (as in a presidential election)—that is, one in which many neurons compete with each other but only one (or a few) win, meaning that it fires more vigorously or in some special way, while all the others are forced to fire more slowly or not at all.

This can be done rather easily in artificial neural networks by making each neuron put out excitation while at the same time inhibiting all the other competing neurons. The one that is the most active is likely to suppress all its rivals (as in an election!). For real neurons this is not so simple, since in most cases a single neuron can put out only excitation or inhibition, but not both. There are various tricks to avoid this difficulty; for example, by making all the excitatory neurons stimulate an inhibitory one, which in turn inhibits all the excitatory ones. The neuron that can triumph best over this general inhibition is then likely to emerge the winner. It requires some skill to design a neural network that will perform a winner-take-all operation in a satisfactory way, but it can be done, especially if several co-winners are allowed rather than a single neuron.

There seems no reason why Nature should not have evolved this kind of mechanism. The problem is how to spot exactly where such operations are happening in the brain. So far we do not know enough about the highly complex local circuitry in or near the cortex to help us much, but this may change as our knowledge increases. It may turn out that the neural interactions in the cortex are so complicated that no simple mechanism is involved, but there is always the chance that such a key process uses some special neural device. All we can do is keep an eye open for promising lines of evidence. The problem is complicated by the fact that awareness may not always demand a decision between two or more choices (as in viewing a Necker cube). In other cases it may be more efficacious to arrive at a compromise between different sources of information, as when judging the distance away of an object in the visual field, using several distinct depth clues. On the other hand, whether one object is in front of another one, and so partially occludes it, does require a decision.

So far our search for the awareness neurons has turned up rather few clues that we can rely on, although it has pointed to several promising lines of investigation. Are there any more approaches we could follow? Can we learn anything useful about visual awareness by studying the neural basis of short-term memory? It seems virtually certain that without any form of very short-term memory we could not be conscious, but how short does it have to be, and what neural mechanisms are involved?

Recall that memory can be of two general types. When you are actively remembering something there must be neurons firing somewhere in your head that symbolize that memory. However, there are many things you can remember, such as the Statue of Liberty or the date of your birth, that you are not actively remembering at one particular moment. Such latent memories will not, in general, require the relevant neurons to be firing. The memory is stored in your brain because the strength of many synaptic connections (and other parameters) were altered, when the memory was laid down, in such a way that the requisite neural activity could be regenerated, given a suitable clue.

Which of these two forms of memory, the active one or the latent one, is involved in the form of very short term memory in which we are interested? It seems highly likely that the active form will often be involved—that is, your immediate memory for an object or event is likely to be based on neurons actively firing. How might this happen? I can think of at least two possible ways.

A neuron, once activated, might continue firing because of some intrinsic property, such as the character of its many ion channels. This firing might continue for a certain time, and then fade away, or it might continue until the neuron got some external signal to make it stop firing. A second rather different mechanism might involve not just the neuron itself but the way it was connected to other neurons. There might be "reverberatory circuits," in which each neuron in a closed ring of neurons excited the next one in the ring, thus keeping the activity going around and around. Both these two possible mechanisms might occur—they are not mutually exclusive.

But could we have, in addition, some latent form of short-term memory? This would mean that the neurons involved would first be stimulated to fire; then they would stop firing but their firing could be rapidly started up again if a sufficiently strong clue was provided to jog the latent memory into activity. But how could this happen unless the first round of firing had left some trace on the system? Are there, perhaps, transient changes in the relevant synaptic strengths (or in other neuronal parameters) that could embody for a short time this brief latent memory? Is there, in fact, any experimental evidence for transient changes at synapses? Their existence, incidentally, was suggested by Christoph von der Malsburg in the rather obscure theoretical paper mentioned earlier.

Unbeknown to Christoph, there was already some experimental evidence for transient synaptic changes. They were originally discovered in the fifties in places very remote from the brain, at neuromuscular junctions where a nerve that activates a muscle contacts that muscle.[5] A little later, similar transient synaptic changes were discovered in the hippocampus (for a review see reference 6). When an axonal spike arrives at a synapse it alters the synapse almost instantaneously, so that its synaptic strength is increased. A quick sequence of spikes can produce a bigger increase. The increase in synaptic strength then decays in complicated ways, partly as fast as 50 milliseconds and partly more slowly, with decay times ranging from fractions of a second to a minute or so. These are the sort of times involved in short-term memory. There is also suggestive evidence that they occur at synapses in the neocortex. They appear to be due mainly to alterations on the input side of a synapse (the presynaptic side) and may involve calcium ions hanging about there and perhaps the movement of synaptic vesicles nearer to the synaptic junction.* Whatever their cause they almost certainly exist. Their size can be quite appreciable.

Unfortunately, little work is now being done on these transient changes, mainly because long-term changes in synaptic strength—a very hot topic at the moment—are easier to study. Nor have they been allowed for in most theoretical work on neural networks. We thus have the curious situation that a phenomenon that may be crucial for con-

---

*If they are solely presynaptic—that is, if they do not depend upon what is happening on the postsynaptic side of the synapse—they are unlikely to be Hebbian, as required by von der Malsburg. Whether there are transient changes that are Hebbian remains to be investigated. The theory of transient non-Hebbian changes has been largely neglected by theorists.

sciousness (and visual awareness in particular) is neglected by both experimentalists and theorists.

These transient changes in synaptic weight could also be important for the temporary maintenance of reverberating circuits. Increasing the relevant synaptic strengths could help the circuit maintain its reverberatory firing.

Perhaps a more difficult problem is how to prevent such sustained firing from spreading too widely and affecting other circuits. There are so many complex pathways in the brain that it may be almost impossible to pin down the exact location of reverberatory circuits, if indeed they exist. Is it possible that this kind of reverberation—the sort associated with active short-term memory—takes place in only one or a few particular places? Is there anything that suggests that such a circuit might be wired to be somewhat isolated from neighboring circuits of the same general kind, so that the memory does not spread in an uncontrolled manner?

Curiously enough, there is one possible circuit that could conceivably be involved in very short term memory. This is the circuit from the thalamus to the type of pyramidal neuron in cortical layer 6, which sends signals back to the same part of the thalamus. Both these thalamic neurons and these cortical neurons have very few axon collaterals that spread sideways, so they probably interact rather little with their neighbors.[7] This may give them the partial isolation just referred to.

These pathways have been mainly studied in cortical area V1 and its connections to the LGN. The forward pathway, from the LGN to the layer 6 pyramidals, appears to be rather weak. The back pathway, from layer 6 to the LGN, has very many axons, perhaps five or ten times as many as the major forward connection from the LGN to layer 4. This is surprising in itself, especially as it has been difficult to discover any function for them. However, most of the experiments on this pathway have been done on anesthetized animals in which the very short term memory mechanism may be weak or absent, thereby making the animal unconscious. Livingstone and Hubel, in the paper described a few pages back, found that the activity of the LGN neurons was reduced in slow-wave sleep. This could well have the effect that while signals could reach cortical area V1 from the LGN (as they found), these signals would not be large enough to sustain any reverberatory activity. There are known neural pathways from the brain stem that can alter the activity of the LGN (and, by extension, the other parts of the thalamus) during slow-wave sleep.

The hypothesis, then, is that these layer 6 neurons are intimately involved in a key aspect of consciousness, the maintenance of reverberatory circuits that embody very short term memory. This is compatible with the general idea, touched on earlier, that it is mainly the lower cortical layers whose activity correlates with consciousness in general, and with visual awareness in particular.

Could there be such reverberatory circuits associated with all cortical areas? In other words, do all cortical areas have pyramidal neurons in their layer 6 that project to some part of the thalamus that itself projects back to those same layer 6 pyramidals? Unfortunately, the facts are not completely clear. Perhaps only the lower and intermediate levels of sensory processing (those that, as it turns out, do have an appreciable layer 4) have the necessary layer 6 reverberatory circuits needed for this form of short-term memory, and thus for conscious awareness of the type Jackendoff has proposed. Possibly a strong input into layer 4 may give certain layer 6 reverberatory circuits more punch. If all this turns out to be true it would make a significant link between the structure of the brain and Jackendoff's hypothesis—an exciting possibility.

Leaving speculation aside, is there any evidence that the sustained firing of neurons is associated with any form of short-term memory? Such experiments have been done by Patricia Goldman-Rakic and her colleagues at Yale,[8] following pioneer work by others.[9] A monkey was trained to fixate on a spot in the center of a TV screen while a target stimulus was presented at a randomly chosen spot elsewhere on the screen. After a delay, during which the target was switched off, the monkey had to move its eyes to where the target had been. The experimenters studied the response of visual neurons in the prefrontal region of the animal's brain. A particular neuron always responded when the target was at a particular place on the screen—other neurons responded to different places. What was striking was that such neurons maintained their firing, often for many seconds *after the target was switched off*, until the monkey made its response. Moreover, if the activity was not maintained, as occasionally happened, the monkey was highly likely to make a mistake. In short, these neurons appeared to be part of a working memory system for visual spatial location.*

*They also showed, using the 2-deoxyglucose technique, that there is more activity during such a task in areas connected to the prefrontal cortex, such as the hippocampal formation, the posterior parietal cortex, and the medial dorsal nucleus of the thalamus.

There are probably other such systems, for other kinds of working memory, elsewhere in the brain. Thus we at least have one example where a short-term memory involves the continued firing of neurons,* although in other cases the evidence is open to doubt.

Notice that a single task is involved and that the monkey may be mentally rehearsing the task during the delay. What would happen neuronally if the monkey had to perform two distinct tasks is not known. Nor is it yet known exactly what neural mechanism keeps these neurons firing. As with the study of attention, we can say that a beginning has been made, but a lot more experimental work will be needed to uncover the neural basis of short-term memory.

---

*Unfortunately, the way these neurons fire does not immediately suggest the existence of reverberatory circuits.

17

Oscillations
and Processing Units

"Prediction is difficult, especially if it concerns the future."

So far I have said very little about possible solutions to the binding problem—how to bind together all the neurons firing to different features of the same object (or event), especially when more than one object is perceived during a single perceptual moment. Binding is important because it seems necessary for at least some types of awareness. In Chapter 14 it was suggested that binding might be achieved by the correlated firing of the neurons concerned. A rather simple kind of correlated firing is that the neurons involved fire together in some sort of rhythm (although rhythms are not essential for correlation). The idealized example in Figure 57 showed neurons firing in bursts every 100 milliseconds—that is, at a frequency of about 10 Hertz. Rhythms with frequencies at or near this are called "α-rhythms." This and other rhythms can be detected in the rather jumbled signals we see in brain waves recorded from the scalp (the EEG). Is there any experimental evidence for correlated firing of sets of individual neurons?

It has been known for some time that correlated firing, in the form of oscillations of firing, occurs in the olfactory system,[1] but it was not till more recently that oscillations were clearly observed in the visual cortex. The most striking results have come from two groups in Germany. Wolf Singer, Charles Gray, and their colleagues at Frankfurt saw oscillations in the cat's visual cortex.[2] These oscillations were in the 35- to 75-Hertz range, often called the "gamma oscillations" or, less accurately, the "40-Hertz oscillations." Reinhard Eckhorn and his colleagues at Marburg independently observed such oscillations.[3] They can be seen especially clearly in an electrode that is made to pick up the "field potentials." Roughly speaking, this displays the constantly changing average activity in the set of neurons near the electrode, rather like picking up the buzz of conversation from a large group of people at a cocktail party.

These experiments are relatively new, and further new results are still coming in. Here I shall only give a very simple, outline description.

As already described, some neurons in the visual cortex fire in a somewhat rhythmical manner when they become active due to a suitable stimulus in the visual field. The average local activity (the field potential) in their vicinity often shows oscillations in the 40-Hertz range. The spikes put out by such a neuron do not occur at random moments, but "on the beat" of the local oscillations (see Fig. 60). A neuron may fire a short burst of two or three spikes on the beat. Sometimes it may not fire at all, but when it does fire it often does so in approximate synchrony with some of its fellow neurons. These oscillations are not especially regular. Their wave form is more like a crude freehand drawing of a wave than a very regular mathematical wave of constant frequency.

If two electrodes are used, not too far apart, Singer and his colleagues often found that the neurons near one electrode, if they fire at all, tend to fire in synchrony with those near the other electrode. Even if the electrodes are as much as 7 millimeters apart the field potentials may oscillate in the same phase, but usually when the moving stimulus used to excite them belongs to one object rather than two,[4] although the experimental evidence to support this last statement is at the moment rather sparse. Additional experiments have shown that neurons in the first visual area can respond to a moving bar by firing rhythmically in phase with those in the corresponding region in the second visual area, showing that synchrony can occur between neurons in different cortical areas[5] and also, in additional experiments, between the two halves of the cortex.[6]

Both German groups suggested that these 40-Hertz oscillations might be the brain's answer to the binding problem. They proposed that the neurons symbolizing all the different attributes of a single

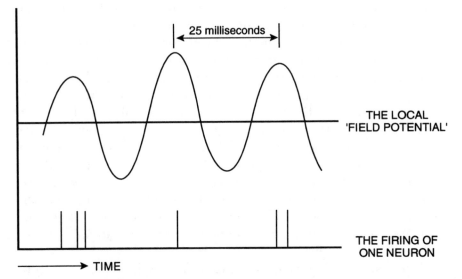

Fig. 60. A simple figure to illustrate how some neurons fire in a 40-Hertz rhythm. (A 40-Hertz oscillation repeats every 25 milliseconds.) The smooth curve represents the local field potential. This is a measure of the average "activity" of many of the neurons in that neighborhood. The short vertical lines show the firing of just one neuron. Notice how, when this neuron fires, it fires "on the beat" of some of its neighbors, represented by the local field potential. (I have reversed the usual sign convention for plotting the field potentials.)

object (shape, color, movement, etc.) would bind these attributes to each other by firing together in synchrony. Koch and I took this idea one stage further by suggesting that this synchronized firing on, or near, the beat of a gamma oscillation (in the 35- to 75-Hertz range) might be the neural correlate of visual awareness.[7] Such behavior would be a special case of the correlated firing suggested by the theorists.

We also suggested that the main function of the attentional mechanism would be to select an object for attention and then help to synchronize the coalition of all the relevant neurons that corresponded to the brain's best interpretation of that part of the visual input. The thalamus, we surmised, was the "organ of attention," some parts of it controlling a spotlight of attention that hopped from one salient object to another in the visual field.

These initial experiments were done when the cat was under a light anesthetic. Under a very deep anesthetic (a barbiturate) no oscillations are seen, but the activity of the neurons under such conditions is in any case much reduced, so this result by itself is not very informative. More recent experiments have been done on alert cats (Charles Gray, personal communication). These also show 40-Hertz oscillations, so the oscillations are not an artifact of the anesthetic. Some recent experiments[8] on cortical area V1 of lightly anesthetized monkeys also show oscillations. They are seen in cortical area MT of alert monkeys when moving bars are used as the visual input[9] but not when patterns of semirandom moving dots are shown.[10] This difference in behavior has not yet been explained. It is more compatible with the oscillations being involved in figure/ground discrimination rather than with visual awareness. They have also been clearly seen by Eberhard Fetz and his colleagues[11] in the motor/somatosensory cortex of an alert monkey, especially when it is doing an intricate manual task that demands its attention.

When the oscillations are seen, they are usually rather transient.[12] How long they last often depends on the duration of the visual signal used. Correlated oscillations between sets of neurons in different places may persist for only a few hundred milliseconds, as some theories might predict. On balance it is hard to believe that our vivid picture of the world really depends entirely on the activities of neurons that are so "noisy" and so difficult to observe.

By now you are likely to be as baffled as the police can be in the early stages of a difficult murder case. There are many possible leads, but so far not one of them has pointed convincingly to the probable solution of the mystery. This is the type of police work that the general public finds hardest to appreciate—the systematic and painstaking following up of many rather weak clues. The same is true of the scientific attack on visual awareness. We all want to know the answer, but this is unlikely to come without carefully checking different trails, many of which are likely to turn out to be misleading or even completely false.

One thing that has emerged from all these considerations is that there may be several forms of visual awareness and, by extension, even more forms of consciousness in general. Is there any way we can map these different forms of visual awareness onto the structure and behavior of the primate visual system?

Recall that I described three possible stages of visual processing: a very transient one, corresponding roughly to Marr's primal sketch; a more lasting and vivid one, related perhaps to his 2½D sketch and to Jackendoff's intermediate level; and a 3D object-centered one that did not correspond to what we actually see but rather to some of the implications of what we see. I see vividly the outlines and visible surfaces of a particular object and this implies that it is a cup, with an inferred 3D shape. In ordinary speech we employ the word *see* for both of these usages. If I say "Do you see that cup over there?," I am using the word *see* in both senses. I could mean just the visible surfaces the cup presents to me, or I could also mean the inferred three-dimensional shape of the whole cup. Notice that the 2½D sketch and the 3D model are both inferences of a sort, in that they have both interpretations of the visual input and that both can be mistaken. Our customary usages of words may not accurately describe the actual activities in our brains.

An idea I call the Processing Postulate states that *each level of visual processing is coordinated by a single thalamic region.** A key question, though one seldom asked, is: What do the cortical areas served by *one* thalamic nucleus have in common?

In the primate visual system we know that the LGN (a part of the thalamus) is related mainly to the first visual area, V1. The other visual regions in the primate thalamus all lie in a large part of it called the "pulvinar" (see Chapter 15), which has a number of distinct subdivisions, some of which may consist of several sub-subdivisions. Is each one of these regions associated with a *single* stage in the visual processing? There are two possibilities. The subdivisions—there are three major ones: the inferior, the lateral, and the medial pulvinar—might each be strongly associated with one of David Marr's stages: the primal sketch, the 2½D sketch, and the 3D model—or something like them. Another possibility is that each of the smaller, more numerous sub-subdivisions is strongly associated with just one of the levels in the Van Essen visual hierarchy (Fig. 52). Of course, both possibilities may have some truth in them.

What do I mean by "strongly associated"? The thalamus sends two sorts of connections to the cortex: one type to layer 4 (or layer 3), and another type that avoids these middle layers and usually projects heavily to layer 1. The first type may be driving connections, while the second is more likely to modulate what is already going in. By strongly associated I mean these driving connections into the middle layers. In this brief account I shall leave aside the other type.

The Processing Postulate, in its simplest form, states that any one

---

*There are many more cortical areas than there are thalamic regions. Since every cortical area is associated with at least one thalamic region, this implies that, in general, one thalamic region will be related to several cortical areas.

cortical area is strongly associated with only one part of the thalamus. This idea is not totally implausible. Cortical area V1 is strongly associated only with the LGN, not with any other part of the thalamus. The features needed to form Marr's primal sketch, or something like it, do appear to be found in V1. The information symbolized there corresponds to rather simple local features, such as orientation in one small part of the visual field. V1 may be the seat of the rather fleeting form of visual awareness that Koch and I postulated.[7] We suggested that this does not need an attentional mechanism. In support of this, experiments have shown[13] that the monkey's attention does not appear to affect the firing of neurons in V1.

Not enough is yet known about the details of the other thalamic connections to decide whether the Processing Postulate is true or not. Is each cortical area (beyond V1) connected strongly with only one part of the pulvinar? If not, how are they connected? More experiments will be needed to answer these questions. It is always possible that some thalamic regions are strongly connected to exactly those cortical areas that are involved in vivid visual awareness.

What about the postulated 3D-model stage? In this case we scarcely know what to look for. The psychologist Irving Biederman believes that such a representation will be based on certain primitive 3D shapes he calls "geons."[14] Some theorists, such as Tomaso Poggio, believe that what we have in the brain is a series of 2D "views" of an object and the ability to interpolate between them.[15] It is even possible that both ideas are correct. Exactly where all this happens in the monkey's brain, if it happens at all, has yet to be determined. This lack of knowledge makes it difficult to assess the Processing Postulate. What looked at first like a beautiful hypothesis seems bogged down in experimental uncertainty.

Nevertheless, the Processing Postulate has a certain appeal. It suggests that we may be using the words *conscious* and *unconscious* for too many somewhat distinct activities. They may have to be replaced by some phrase like "processing unit" or, in some cases, "awareness unit." Each of these units would have its own semiglobal representation, usually covering several cortical areas. Each might have its own characteristic processing time, its own characteristic times for very short term memory (e.g., very short in V1, longer in the higher visual areas), and, most important, *its own particular form of representation*; simple features in V1, 2½D objects in the next higher cortical areas, and so on. *The character of each type of processing unit would depend on the content and organization of that particular representation.* Each particular thalamic region may handle its own form of attention, possibly by allowing neurons in its member set of cortical areas to talk to neurons in the thalamus, which in turn feed back to them, so that in some way their firing is coordinated. There is also the speculative idea (described in Chapter 16) that the thalamo-cortico-thalamic circuit may be intimately involved as a reverberatory circuit for very short term memory.

Of course there are many complex interactions between different cortical areas that are direct (i.e., they do not have to go through the thalamus) as shown by Figure 52. The Processing Postulate does not imply that neural activity flows only one way, from a lower processing unit to a higher one. It will almost certainly flow in several directions.

This does not mean that the thalamus by *itself* can produce all the different forms of awareness. Awareness requires the activity of the various cortical areas as well as the thalamus, just as a conductor needs the orchestra to produce music.* The minimum message, then, is that if you are interested in visual awareness, or in other aspects of consciousness for that matter, the thalamus should not be neglected. One might brush aside the poor LGN and say that it is only a relay, but a student of the visual system should ask, "Why is the pulvinar there at all?" It is not a trivially small region of the brain and indeed has become larger in primate evolution. It is likely to have *some* important function, but what is it? The Processing Postulate, while vague about details, does suggest one such possibility.

The idea that the thalamus is a key player in consciousness is not a new one. It was suggested much earlier[16] by Wilder Penfield.† In a recent publication[18] James Newman and Bernard Baars have extended the latter's ideas (discussed briefly in Chapter 2) and now suggest that the information broadcast to their postulated global workspace comes from a certain set of thalamic regions called the "intralaminar nuclei." One of these nuclei—the nucleus centralis—is closely associated with the visual system. These nuclei project mainly to an important part of the brain, the corpus striatum, although they also project to a lesser extent to many cortical regions. The striatum is intimately connected with the motor system, but some parts of it may be concerned with more cognitive matters. It is one of the parts of the brain that is affected in Parkinson's disease.

*Does the orchestra need the conductor? This loose analogy should not be pursued too far, but it may be significant that a small group of musicians does not need a conductor, while a very large one usually does.

†More recently it has been put forward by David Mumford, the Harvard mathematician,[17] and also in preprints of papers sent to me by Quanfeng Wu.

Exactly what information each intralaminar nucleus sends out remains to be seen.* Newman and Baars also place great emphasis on the reticular nucleus of the thalamus (described in Chapter 10). They believe it may be involved in gating attention as I myself once speculated.[19] At the moment, it is unclear whether the reticular nuclei can exert the required degree of selectivity over the thalamus. Perhaps its only function is to act as an overall controller of thalamic and cortical activity during brain states such as sleeping and waking. If the thalamus is indeed the key to consciousness, the reticular nucleus is likely to play some part in the control of consciousness.

There is one other brain region that must be mentioned briefly. This is the claustrum,[20] a thin sheet of neurons lying next to the lower cortical layers near a part of the cortex called the "insula." It appears to be a satellite of the cortex since its input comes mainly from the cortex and most of its output goes back to the cortex. It receives an input from many cortical areas and may send connections back to all of them. Some but not all the visual areas of the cortex project to one part of it and (in the cat) form a single retinotopic map there. There may be some overlap of this visual input with other claustral inputs. Very little work seems to have been done on the monkey claustrum in these last few years, so some of the above statements may be somewhat inaccurate. (For example, it is possible there are *two* visual maps there.)

Its function is unknown. Why should all this information be brought together in one thin sheet? One would suspect that the claustrum has some kind of global function, but what that might be nobody knows. Even though it is rather a small region of the brain, it should not be totally overlooked.

It may well be that there is a hierarchy of processing units, in the sense that some of them may exert some sort of global control over the others. Sets of neurons that project very widely over the cortex, such as the claustrum and the intralaminar nuclei of the thalamus, could well play such a role.

Reviewing the last two chapters, it can be seen that there is no lack of plausible ideas and feasible experiments. What is disappointing is that, at the moment of writing, there does not appear to be one set of ideas that click together in a convincing way to make a detailed neural

---

*It has been suggested that the nucleus centralis is mainly involved in the control of gaze.

hypothesis that has the smell of being correct. If you think I appear to be groping my way through the jungle you are quite right. Work at the research front is often like that. I do feel, however, that I now have a better understanding of what the key problems are than I did ten years ago. At times I even persuade myself that I can glimpse some of the answers, but this is a common delusion experienced by anyone who dwells too long on a single problem. We have yet to break through to the high ground so that, even if the way is long and difficult, we can see the best direction to explore.

In spite of all such uncertainties, is it possible, after considering all these very diverse facts and speculations, to sketch some overall scheme, however tentative, that might act as a rough guide through the jungle ahead of us? Let me throw caution to the winds and outline one possible model. Reality may be more complex than this model but it is unlikely to be much simpler.

Consciousness is associated with certain neural activities. A plausible model could start with the idea that this activity is largely in the lower cortical layer (layers 5 and 6). This activity expresses the local (transient) results of "computations" taking place mainly in other cortical layers.

Not all cortical neurons in the lower cortical layers can express consciousness. The most likely types are some of the large "bursty" pyramidal cells in layer 5, such as those that project right out of the cortical system.

This special lower-layer activity will not reach consciousness unless sustained by some form of very short term memory. It is plausible that this needs an effective reverberatory circuit from cortical layer 6 to the thalamus and back again to cortical layers 4 and 6. If this is lacking, or if layer 4 is too small, it may not be possible to sustain these reverberations. For this reason only some cortical areas will express consciousness.

A processing unit (only some of which are associated with consciousness) is a set of cortical areas* at the same level in the visual hierarchy, each projecting into each other's layer 4. Each set of such cortical areas is strongly connected to just one small region of the thalamus. Such a region coordinates the activities of its associated cortical areas by synchronizing their firing.

The thalamus is intimately involved in attentional mechanisms. Special binding, where needed for operations such as object tagging (especially figure/ground discrimination), takes the form of coordinated firing, often with rhythms in the 40-Hertz range.

*Some sets may have only one member, as in the case of V1.

The regions involved in consciousness can influence, not necessarily directly, parts of the voluntary motor system. (There may be *uncon*-scious operations—"thoughts"—in between.)

To repeat: Consciousness depends crucially on thalamic connections with the cortex. It exists only if certain cortical areas have reverberatory circuits (involving cortical layers 4 and 6) that project strongly enough to produce significant reverberations.

So much for a plausible model. I hope nobody will call it the Crick (or the Crick-Koch) Theory of Consciousness. While writing it down, my mind was constantly assailed by reservations and qualifications. If anyone else produced it, I would unhesitatingly condemn it as a house of cards. Touch it, and it collapses. This is because it has been carpentered together, with not enough crucial experimental evidence to support its various parts. Its only virtue is that it may prod scientists and philosophers to think about these problems in neural terms, and so accelerate the experimental attack on consciousness.

What of more philosophical questions? I believe that when the neural basis of consciousness is thoroughly understood this knowledge will suggest answers to two major questions: What is the *general* nature of consciousness, so that we can talk sensibly about the nature of consciousness in other animals, and also in man-made machines, such as computers? What advantage does consciousness give an organism, so that we can see why it has evolved? It may turn out that visual awareness has arisen because the detailed information in it needs to be sent to several distinct places in the brain. It may be more efficient to make this information explicit, once and for all, rather than relaying the information, in a tacit form, along a number of distinct, parallel pathways. Having one single explicit representation would also keep one part of the brain from using one interpretation of the visual scene while another part used a rather different one. When the information needs to be sent to only one place then, with experience, it may be routed there without consciousness being involved.

What may prove difficult or impossible to establish are the details of the subjective nature of consciousness, since this may depend upon the exact symbolism employed by each conscious organism. This symbolism may be impossible to convey to another organism in a direct manner until and unless we can hook two brains together in a sufficiently precise and detailed way, and even if we could do this it might raise problems of its own. But without understanding the neural correlate of consciousness I do not believe that any of these questions can be answered in a manner that all thoughtful people would accept.

What could delay the experimental attack is the rather conservative attitude taken by many of the scientists who actively work on the brain and on vision in particular. Let me say a special word to them.

They will see only too clearly the many complex issues that I have slurred over or ignored. They should not use these errors and omissions to excuse themselves from confronting the broad message of this book. It will no longer do to study in detail some particular aspect of vision and at the same time to ignore the overall question: What happens in the brain to make us see? A layman would regard such an attitude as excessively narrow-minded and he would be right. As I have tried to show, the problem of visual awareness is approachable now, both experimentally and theoretically. What is more, by actively confronting the problem, one begins to think in new ways; to ask for information (e. g., on dynamic parameters or short-term memory) that had previously appeared to be irrelevant or of little interest. I hope before long every laboratory working on the visual system of man and the other vertebrates will have a large sign posted on its walls, reading:

CONSCIOUSNESS
NOW

18

Dr. Crick's
Sunday Morning Service

"Our own subjective inner life, including sensory experiences, feelings, thoughts, volitional choices, is what really
matters to us as human beings."

—Benjamin Libet

The time has come to take stock of where the problem stands.
We have seen how complex the visual system is and how visual
information is processed in a semihierarchical manner that is
only partly understood. I have sketched a few ideas about the neural
basis of visual awareness and outlined a few experiments that might
help to unscramble its mechanisms. Clearly we have not solved the
problem, so what has been achieved?

What Christof Koch and I are trying to do is to persuade people,
and especially those scientists intimately involved with the brain, that
now is the time to take the problem of consciousness seriously. We
suspect that it is our general approach that will be found useful, rather
than our detailed suggestions. The speculations discussed in this book
are not a fully worked out, coherent set of ideas. Rather, they constitute work in progress. I believe that the correct way to conceptualize
consciousness has not yet been discovered and that we are merely
groping our way toward it. This is one reason why experimental evi-

dence is important. New results may suggest both new ideas and also alert us to errors in old conceptions.

Philosophers are right in trying to discover better ways of looking at the problem and in suggesting fallacies in our present thinking. That they have made so little real progress is because they are looking at the system from outside. This makes them use the wrong idiom. It is essential to think in terms of neurons, both their internal components and the intricate and unexpected ways they interact together. Eventually, when we truly understand how the brain works, we may be able to give approximate high-level accounts of our perceptions, our thinking, and our behavior. This will help us to grasp the overall performance of our brains in a more correct and more compact manner, and will replace the fuzzy folk notions we have today.

Many philosophers and psychologists believe it is premature to think about neurons now. But just the contrary is the case. It is premature to try to describe how the brain really works using just a black-box approach, especially when it is couched in the language of common words or the language of a digital programmable computer. *The language of the brain is based on neurons.* To understand the brain you must understand neurons and especially how vast numbers of them act together in parallel.

The reader might accept all this but could well complain that I have talked all around the topic of consciousness, with more speculation than hard facts, and have avoided what, in the long run, is the most puzzling problem of all. I have said almost nothing about qualia—the redness of red—except to brush it to one side and hope for the best. In short, why is the Astonishing Hypothesis so astonishing? Is there some aspect of the brain's structure and behavior that might suggest why it is so difficult for people to conceive of awareness in neural terms?

I think there is. I have described the general workings of an intricate machine—the brain—that handles an immense amount of information all at once, in one perceptual moment. Much of the content of this rich body of coherent information is constantly changing, yet the machine manages to keep various running records of what it has just been doing. We have no experience (apart from the very limited view provided by our own introspection) of any machine with all these properties, so it is not surprising that the results of that introspection appear so odd. Johnson-Laird has made a similar point (quoted in Chapter 14). If we could build machines that had these astonishing characteristics, and could follow exactly how they worked, we might find it easier to grasp the workings of the human brain. The mysterious aspects of consciousness might disappear, just as the mysterious aspects of embryology have largely disappeared now that we know about the capabilities of DNA, RNA, and protein.

This leads to the obvious question: Will we be able, in the future, to build such machines and, if we did, would they appear to possess consciousness? I believe that in the long run this may be possible, although there may turn out to be technical limitations that are almost impossible to overcome. In the near future I suspect that any machines we can construct are likely to be very simple in their capabilities compared to the human brain. Because of this they will probably appear to have only a very limited form of consciousness. Perhaps they will be more like the brain of a frog or even that of a humble fruit fly. Until we understand what makes us conscious, we are not likely to be able to design the right sort of artificial machine or to arrive at firm conclusions about consciousness in lower animals.

It is important to emphasize that the Astonishing Hypothesis is a hypothesis. What we already know is certainly enough to make it plausible, but it is not enough to make it as certain as science has done for many new ideas about the nature of the world, and about physics and chemistry in particular. Other hypotheses about man's nature, especially those based on religious beliefs, are based on evidence that is even more flimsy but this is not in itself a decisive argument against them. Only scientific certainty (with all its limitations) can in the long run rid us of the superstitions of our ancestors.

A critic could argue that, whatever scientists may say, they really do believe in the Astonishing Hypothesis. There is a restricted sense in which this is true. You cannot successfully pursue a difficult program of scientific research without some preconceived ideas to guide you. Thus, loosely speaking, you "believe'" in such ideas. But to a scientist these are only provisional beliefs. He does not have a blind faith in them. On the contrary, he knows that he may, on occasion, make real progress by disproving one of his cherished ideas. That scientists have a preconceived bias toward scientific explanations I would not deny. This is justified, not just because it bolsters their morale but mainly because science in the past few centuries has been so spectacularly successful.

The next thing to stress is that the study of consciousness is a scientific problem. Science is not separated from it by some insurmountable barrier. If there is any lesson to be learned from this book it is that we can now see ways of approaching the problem experimentally. There is no justification for the view that only philosophers can deal with it.* Philosophers have had such a poor record over the last two thousand years that they would do better to show a certain modesty rather than the lofty superiority that they usually display. Our tentative ideas about the workings of the brain will almost certainly need clarification and extension. I hope that more philosophers will learn enough about the brain to suggest ideas about how it works, but they must also learn how to abandon their pet theories when the scientific evidence goes against them or they will only expose themselves to ridicule.

The record of religious beliefs in explaining scientific phenomena has been so poor in the past that there is little reason to believe that the conventional religions will do much better in the future. It is certainly possible that there may be aspects of consciousness, such as qualia, that science will not be able to explain. We have learned to live with such limitations in the past (e.g., limitations of quantum mechanics) and we may have to live with them again. This does not necessarily mean that we shall be driven to embrace traditional religious beliefs. Not only do the beliefs of most popular religions contradict each other but, by scientific standards, they are based on evidence so flimsy that only an act of blind faith can make them acceptable. If the members of a church really believe in a life after death, why do they not conduct sound experiments to establish it? They may not succeed but at least they could try. History has shown that mysteries which the churches thought only they could explain (e.g., the age of the earth) have yielded to a concerted scientific attack. Moreover, the true answers are usually far from those of conventional religions. If revealed religions have revealed anything it is that they are usually wrong. The case for a scientific attack on the problem of consciousness is extremely strong. The only doubts are how to go about it, and when. What I am urging is that we should pursue it now.

There are, of course, educated people who believe that the Astonishing Hypothesis is so plausible that it should not be called astonishing. I have touched on this briefly in the first chapter. I suspect that such people have often not seen the full implications of the hypothesis. I myself find it difficult at times to avoid the idea of a homunculus. One slips into it so easily. The Astonishing Hypothesis

---

*It has been unkindly said that a philosopher is too often a person who prefers imaginary experiments to real experiments and thinks that an explanation of a phenomenon in everyday words is all that is needed.

states that *all* aspects of the brain's behavior are due to the activities of neurons. It will not do to explain all the various complex stages of visual processing in terms of neurons and then carelessly assume that some aspect of the act of seeing does not need an explanation because it is what "I" do naturally. For example, you cannot be aware of a defect in your brain unless there are neurons whose firing symbolizes that defect. There is no separate "I" who can recognize the defect independent of neural firing. In the same way, you do not normally know where something is happening in your brain because there are no neurons in your brain whose firing symbolizes where they or any other neurons in your brain are situated.

Many of my readers might justifiably complain that what has been discussed in this book has very little to do with the human soul as they understand it. Nothing has been said about that most human of capabilities—language—nor about how we do mathematics, or problem solving in general. Even for the visual system I have hardly mentioned visual imagination or our aesthetic responses to pictures, sculpture, architecture, and so on. There is not a word about the real pleasure we get from interacting with Nature. Topics such as self-awareness, religious experiences (which can be real enough, even if the customary explanations of them are false), to say nothing of falling in love, have been completely ignored. A religious person might aver that what is most important to him is his relationship with God. What can science possibly say about that?

Such criticisms are perfectly valid at the moment, but making them in this context would show a lack of appreciation of the methods of science. Koch and I chose to consider the visual system because we felt that, of all possible choices, it would yield most easily to an experimental attack. The book shows clearly that while such an attack will not be easy, it does appear to have some chance of success. Our other assumption was that, once the visual system is fully understood, the more fascinating aspects of the "soul" will be much easier to study. Only time will show if these judgments are correct. New methods and new ideas might make other approaches more attractive. The aim of science is to explain *all* aspects of the behavior of our brains, including those of musicians, mystics, and mathematicians. I do not contend that this will happen quickly. I do believe that, if we press the attack, this understanding is likely to be reached some day, perhaps some time during the twenty-first century. The sooner we start, the sooner we shall be led to a clear understanding of our true nature.

*   *   *

Of course, there are people who say that they do not wish to know how their minds work. They believe that to understand Nature is to diminish her, since it removes the mystery and thus the natural awe that we feel when we are confronted with things that impress us but about which we know very little. They prefer the myths of the past even when they are in clear contradiction to the science of the present. I do not share this view. It seems to me that the modern picture of the universe—far, far older and bigger than our ancestors imagined, and full of marvelous and unexpected objects, such as rapidly rotating neutron stars—makes our earlier picture of an earth-centered world seem far too cozy and provincial. This new knowledge has not diminished our sense of awe but increased it immeasurably. The same is true of our detailed biological knowledge of the structure of plants and animals, and of our own bodies in particular. The psalmist said, "I am fearfully and wonderfully made," but he had only a very indirect glimpse of the delicate and sophisticated nature of our molecular construction. The processes of evolution have produced wonders of which our civilized ancestors had no knowledge at all. The mechanism of DNA replication, while basically unbelievably simple and elegant, has been elaborated by evolution to one of great complexity and precision. One must be dull of soul indeed to read about it and not feel how marvelous it is. To say that our behavior is based on a vast, interacting assembly of neurons should not diminish our view of ourselves but enlarge it tremendously.

It has been reported that a religious leader, shown a large drawing of a single neuron, exclaimed, "So that is what the brain is like!" But it is not the single neuron, wonderful though it is as an elaborate and well-organized piece of molecular machinery, that is built in our image. The true description of us is the complex, ever-changing pattern of interactions of billions of them, connected together in ways that, in their details, are unique to each one of us. The abbreviated and approximate shorthand that we employ every day to describe human behavior is a smudged caricature of our true selves. "What a piece of work is a man!" said Shakespeare. Had he been living today he might have given us the poetry we so sorely need to celebrate all these remarkable discoveries.

It is unlikely that the Astonishing Hypothesis, if it turns out to be true, will be universally accepted unless it can be presented in such a way that it appeals to people's imagination and satisfies their need for a coherent view of the world and themselves in terms they can easily understand. It is ironic that while science aims at exactly such a unified view, many people find much of our present scientific knowledge too inhuman and too difficult to understand.

This is not surprising since most of it is in the fields of physics, chemistry, and their related disciplines, such as astronomy, all of which are somewhat removed from the daily lives of most people. In the future this may change. We can hope to understand more precisely the mechanisms of such mental activities as intuition, creativity, and aesthetic pleasure, and in so doing grasp them more clearly and, it is to be hoped, enjoy them more. Free Will (see the Postscript) may no longer be a mystery. That is why the words *nothing but* in our hypothesis can be misleading if understood in too naive a way. Our wonder and appreciation will come from our insights into the marvelous complexities of our brains, complexities we can only glimpse today.

While we may not be able to deduce human values solely from scientific facts, it is idle to pretend that scientific knowledge (or *unscientific* knowledge, for that matter) has no influence on how we form our values. To construct a New System of the World we need both inspiration and imagination, but imagination building on flawed foundations will, in the long run, fail to satisfy. Dream as we may, reality knocks relentlessly at the door. Even if perceived reality is largely a construct of our brains, it has to chime with the real world or eventually we grow dissatisfied with it.

If the scientific facts are sufficiently striking and well established, and if they support the Astonishing Hypothesis, then it will be possible to argue that the idea that man has a disembodied soul is as unnecessary as the old idea that there was a Life Force. This is in head-on contradiction to the religious beliefs of billions of human beings alive today. How will such a radical challenge be received?

It would be comforting to believe that most people would be so convinced by the experimental evidence that they would immediately change their views. Unfortunately, history suggests otherwise. The age of the earth is now established beyond any reasonable doubt as very great, yet in the United States millions of Fundamentalists still stoutly defend the naive view that it is relatively short, an opinion deduced from reading the Christian Bible too literally. They also usually deny that animals and plants have evolved and changed radically over such long periods, although this is equally well established. This gives one little confidence that what they have to say about the process of natural selection is likely to be unbiased, since their views are predetermined by a slavish adherence to religious dogmas.

There seem to me to be several root causes of this obstinate clinging to outmoded ideas. General ideas, especially moral ones, impressed on us at an early age often become deeply embedded in our brains. It can be very difficult to change them. This may help to explain why religious beliefs persist from generation to generation, but how did such ideas originate in the first place, and why do they so often turn out to be incorrect?

One factor is our very basic need for overall explanations of the nature of the world and of ourselves. The various religions provide such explanations and in terms the average person finds easy to relate to. It should always be remembered that our brains largely developed during the period when humans were hunter-gatherers. There was strong selective pressure for cooperation within small groups of people and also for hostility to neighboring, competing tribes. Even in this century, in the forests of the Amazon, the major cause of death among the competing tribes in the remote parts of Ecuador is from spear wounds inflicted by members of rival tribes. Under such circumstances a shared set of overall beliefs strengthens the bond between tribal members. It is more than likely that the need for them was built into our brains by evolution. Our highly developed brains, after all, were not evolved under the pressure of discovering scientific truths but only to enable us to be clever enough to survive and leave descendants.

From this point of view there is no need for these shared beliefs to be completely correct, provided people can believe in them. The single most characteristic human ability is that we can handle a complex language fluently. We can use words to denote not only objects and events in the outside world but also more abstract concepts. This ability leads to another strikingly human characteristic, one that is seldom mentioned: our almost limitless capacity for self-deception. The very nature of our brains—evolved to guess the most plausible interpretations of the limited evidence available—makes it almost inevitable that, without the discipline of scientific research, we shall often jump to wrong conclusions, especially about rather abstract matters.

How it will all turn out remains to be seen. The Astonishing Hypothesis may be proved correct. Alternatively, some view closer to the religious one may become more plausible. There is always a third possibility: that the facts support a new, alternative way of looking at the mind-brain problem that is significantly different from the rather crude materialistic view many neuroscientists hold today and also from the religious point of view. Only time, and much further scientific work, will enable us to decide. Whatever the answer, the only sensible way to arrive at it is through detailed scientific research. All other approaches are little more than whistling to keep our courage up. Man is endowed with a relentless curiosity about the world. We cannot be satisfied forever by the guesses of yesterday, however much the charms of tradition and ritual may, for a time, lull our doubts about their validity. We must hammer away until we have forged a clear and valid picture not only of this vast universe in which we live but also of our very selves.

A Postscript on Free Will

"Consciousness the Will informing . . ."
—Thomas Hardy

Free Will is, in many ways, a somewhat old-fashioned subject. Most people take it for granted, since they feel that usually they are free to act as they please. While lawyers and theologians may have to confront it, philosophers, by and large, have ceased to take much interest in the topic. And it is almost never referred to by psychologists and neuroscientsts. A few physicists and other scientists who worry about quantum indeterminacy sometimes wonder whether the uncertainty principle lies at the bottom of Free Will.

I myself had paid little attention to Free Will until 1986, when I received a letter from an old friend, Luis Rinaldini, an Argentinian cell biologist whom I had first known in Cambridge in the late forties. Luis and his wife now live in Mendoza, a provincial city in Argentina, near the Andes. He was coming to the United States on a visit and wanted to meet and talk over some of his ideas. When we got together in New York he told me that he and a group of friends had formed a discussion group in Mendoza, and that he had become interested in the topic of Free Will. He subsequently wrote to me on the subject in more detail.

Up to that point I was not aware I had a theory of Free Will, but from what he wrote about it I could see that my ideas differed somewhat from his. I there and then wrote out, very briefly, what I discovered I believed and sent it to him. The text occupied less than thirty lines. I showed it to the philosopher Patricia Churchland, partly for some reassurance that I wasn't being totally silly. She helpfully clarified the wording and added an extra point, telling me that my ideas seemed plausible to her. What follows is a slightly expanded version of what I wrote to Luis.

My first assumption was that part of one's brain is concerned with making plans for future actions, without necessarily carrying them out. I also assumed that one can be conscious of such plans—that is, that they are subject at least to immediate recall.

My second assumption was that one is not conscious of the "computations" done by this part of the brain but only of the "decisions" it makes—that is, its plans. Of course, these computations will depend on the structure of that part of the brain (derived partly epigenetically and partly from past experience) and on its current inputs from other parts of the brain.

My third assumption was that the decision to act on one plan or another is also subject to the same limitations. In other words, one has immediate recall of what is decided but not of the computations that went into the decision, even though one may be aware of a plan to move.*

Then, such a machine (this was the word I used in my letter) will appear to itself to have Free Will, provided it can personify its behavior—that is, it has an image of "itself."

The actual cause of the decision may be clear cut (Pat's addition), or it may be deterministic but chaotic—that is, a very small perturbation may make a big difference to the end result. This would give the appearance of the Will being "free" since it would make the outcome essentially unpredictable. Of course, conscious activities may also influence the decision mechanism (Pat's addition).

Such a machine can attempt to explain to itself why it made a certain choice (by using introspection). Sometimes it may reach the correct conclusion. At other times it will either not know or, more likely, will confabulate, because it has no conscious knowledge of the "reason" for the choice. This implies that there must be a mechanism for confabulation, meaning that given a certain amount of evidence, which may or may not be misleading, part of the brain will jump to the simplest conclusion. As we have seen, this can happen all too easily.

This concluded my Theory of Free Will. It obviously depends upon understanding what consciousness is about (the main topic of this book), how the brain plans (and carries out) actions, how we confabulate, and so on. I doubt if there is anything really novel in all this, although some of the details may not have been included in previous explanations.

---

*Professor Piergiorgio Odifreddi has pointed out to me that one should also assume that there is agreement between decisions and the resulting behavior.

* * *

And there I was content to let the matter rest. I met Luis in New York, and subsequently he came to La Jolla, California, on a visit. He was also able to discuss the problem with Paul Churchland (husband of Patricia Churchland). I had not intended to ponder more on the topic but once my interest had been aroused I found myself thinking about it from time to time.

Where, I wondered, might Free Will be located in the brain? Obviously it involves interactions between several parts of the brain but it was not unreasonable to think that one part of the cortex might be especially concerned. One might expect that this received output from the higher levels of the sensory systems and involved, or fed into, the higher, planning levels of the motor system.

At this point I happened to stumble across an account by Antonio Damasio and his colleagues of a woman with certain brain damage.[1] After the damage, she appeared very unresponsive. She lay quietly in bed with an alert expression. She could follow people with her eyes but did not speak spontaneously. She gave no verbal reply to any questions put to her even though it appeared she understood them because of the way she nodded in reply. She could repeat words and sentences but only very slowly. In short, her few reactions were very limited and rather stereotyped.

After a month she had largely recovered. She said she had not been upset because she had previously been unable to communicate. She had been able to follow conversations but she had not talked because she had "nothing to say." Her mind had been "empty." I immediately thought "she'd lost her Will"—where was the damage? It turned out to be somewhere in or near* a region called the "anterior cingulate sulcus," next to Brodmann's area 24. This is on the inside surface—the one you'd see if the brain were cut in half—toward the front (hence anterior) and near the top. I was delighted to learn that this was indeed a region that received many inputs from the higher sensory regions and was at or near the higher levels of the motor system.

Terry Sejnowski's group at the Salk Institute has an informal tea on most afternoons of the week. These teas are ideal occasions to discuss the latest experimental results, throw out new ideas, or just gossip about science, politics, or the news in general. I went over to tea one day and announced to Pat Churchland and Terry Sejnowski that the seat of the Will had been discovered! It was at or near the anterior cin-

---

*There was also damage to the adjacent supplementary motor area.

gulate. When I discussed the matter with Antonio Damasio, I found that he also had arrived at the same idea. He filled me in on some of the anatomical connections of that region of the brain. It has strong connections with the corresponding area on the other side of the brain—you normally have only a single Will at any one moment, although, as we have seen, split brains can have two (see Chapter 12). Moreover, that region on one side projects strongly to the corpus striatum (an important part of the motor system) on *both* sides of the brain, which is what you might expect from a single Will. It certainly looked rather promising.

Some time later I was reading a paper by Michael Posner in which he mentioned a curious condition, produced by a particular kind of brain damage, known as the "alien hand" syndrome. A patient's left hand, for example, may make movements—usually rather simple and stereotyped movements—that the patient denies he himself is responsible for.[2] For example, the hand may spontaneously grasp some object put near it. In some cases the patient is unable to get the hand to let go and has to use his right hand to detach the left hand from the object. One patient found that he could not make his "alien" hand let go by his own willpower, but could make it release its grasp by saying "let go" in a loud voice!

And where was the damage? Again, in or near the anterior cingulate sulcus (on the right side, for an alien left hand) but *also* in that part of the corpus callosum, so that the region on the left could not give the instructions to the left hand that the damaged region on the right was unable to give. Moreover, as mentioned in Chapter 8, the anterior cingulate is active in certain selection processes, as shown by the increased blood flow there.

So perhaps this aspect of the idea is novel.* Free Will is located in or near the anterior cingulate sulcus. In practice, things are likely to be more complicated. Other areas in the front of the brain may also be involved. What is needed are more experiments on animals, the careful examination of more cases of the "alien hand" and related conditions, and, above all, a detailed neurobiological understanding of visual awareness and, from that, of other forms of consciousness. And that is why this suggestion is appended to this book.

---

*Sir John Eccles had previously suggested[3] that an area near area 24, the supplementary motor area, might be the seat of Free Will.

Units of Length, Time, and Frequency

In discussing nerve cells, the most useful unit of length is the micron. This is sometimes written μm (short for micrometer) and referred to colloquially as a "μ" (pronounced "mew").

$$1 \text{ micron} = 1 \text{ thousandth of a millimeter, and so}$$
$$25.4 \text{ microns} = 1 \text{ thousandth of an inch}$$
$$1 \text{ micron} = 10^{-6} \text{ meters}$$

The diameter of the cell body of a typical neuron is about 10 or 20 microns. The wavelengths of visible light are in the region of ½ micron. For discussing atoms, the preferred unit is the Ångstrom unit.

$$10,000 \text{ Å} = 1 \text{ micron}$$
$$10 \text{ Å} = 1 \text{ nanometer}$$

The distance between adjacent atoms in an organic molecule is mostly about 1 or 2 Å. An average-sized protein might have a diameter of 50 Å, but many larger and smaller ones exist.

TIME

For describing the behavior of neurons, the most useful unit of time is the millisecond.

$$1 \text{ millisecond} = 1 \text{ thousandth of a second}$$
$$1 \text{ microsecond} = 1 \text{ thousandth of a millisecond}$$

FREQUENCY

1 Hertz = 1 event or cycle per second
Middle C is about 260 Hertz

Glossary

Acetylcholine A small chemical neurotransmitter. It is excreted by motor nerves to excite skeletal muscles. It is also used in parts of the brain.

Achromatopsia The inability to see colors while seeing satisfactorily in black and white. Usually caused by damage to a particular part of the brain.

Action potential The all-or-none electrical pulse that travels down an axon, normally from the cell body to the many synapses at the far end of the axon.

Algorithm A rule for solving a particular problem. In many cases the rule consists of applying a special set of steps over and over again, as, for example, in long division. There are many different types of algorithm.

Alien hand A condition, usually produced by brain damage, in which one of an individual's hands makes simple movements that he denies he willed.

Ames room A distorted room named after the psychologist Adelbert Ames. Viewed with one eye through a fixed hole in one of its walls, it produces false perspective. See Figure 14.

Annulus A circular disk with a concentric circular hole inside it. In everyday terms, a two-dimensional doughnut.

Anterior commissure A tract of nerve fibers (axons) toward the front of the brain that links various brain regions with some of those on the opposite side of the head.

Anton's syndrome A rare medical condition that is the result of cortical brain damage. The individual is truly blind but denies that he cannot see. Also called "blindness denial."

Aperture problem The problem of how to detect the true movement of a straight, featureless line viewed only through a small, usually circular, aperture. See page 153.

Archicortex See HIPPOCAMPUS; CEREBRAL CORTEX.

Artificial Intelligence (A.I.) Usually written (and pronounced) A.I. The study of how to make computers behave intelligently. It is done both to improve computer technology and to help us understand how the brain works.

The Astonishing Hypothesis The hypothesis that a person's mental activities are entirely due to the behavior of nerve cells, glial cells, and the atoms, ions, and molecules that make them up and influence them. The theme of this book.

Attention Concentration on a particular stimulus, sensation, or thought while excluding others. A broad term that probably covers more than one type of brain mechanism.

Axon The output cable of a neuron. A neuron usually has only one axon, although it often branches extensively.

Backprop Short for the "back propagation of errors." An algorithm for adjusting the weights in a supervised, multilayered network, especially for simple feed-forward networks with several layers. See page 188ff.

Basket cell In the cerebral cortex, a type of inhibitory nerve cell, often with a rather long axon, that tends to make multiple contacts on or near the cell bodies of other neurons.

Behaviorism A movement in psychology that believed mental events should be ignored and that only stimuli and the responses to them were worth studying.

Binocular rivalry When very different images are presented to each eye, the brain suppresses first one and then the other, in alternation, rather than trying to combine the two into a single percept. See pages 219–222.

Blindsight Produced by brain damage. The ability to respond to certain simple visual signals while denying that one can see them. See page 171.

Blind spot A region in the retina having no photoreceptors.

Brain waves An everyday term for broad electrical activity of the brain, usually recorded by electroencephalography using electrodes attached to the skull.

Broca's area An area in the dominant hemisphere, toward the front of the brain, concerned with certain aspects of language. Damage to it produces a characteristic form of aphasia. Broca's area is almost certainly a group of distinct cortical areas in the modern sense of the term.

Cerebellum A large brain structure, located behind the brain stem at the back of the head. Its large cortex is a relatively simple structure. It is believed to be mainly involved in some way with the fine control of movement.

Cerebral cortex Often called simply "the cortex." A pair of large folded sheets of nervous tissue, one on either side of the top of the head, see Figure 23. It is sometimes subdivided into three main regions, called the neocortex (the largest part in primates), the paleocortex, and the archicortex.

Chandelier cell A type of inhibitory neuron in the cerebral cortex whose axons form multiple synapses with the initial segment of the axons of many pyramidal cells.

Cheshire Cat effect An example of binocular rivalry. A moving object, such as a hand, seen with one eye can make invisible an object, or part of an object, seen (in the same part of the visual field) by the other eye. If the object seen is a smiling face it sometimes happens that most of the face disappears while the smiling lips remain. Hence the name, after the Cheshire Cat in *Alice in Wonderland*.

Cingulate A part of the cerebral cortex on its inner (medial) surface. The anterior cingulate is in the front of the brain.

Cognitive science Any discipline that studies cognition scientifically. Its main branches are linguistics, cognitive psychology, and Artificial Intelligence. In Stuart Sutherland's opinion, "the expression allows workers who are not scientists to claim that they are." He adds, "...Cognitive scientists rarely pay much attention to the nervous system."

Cone A special type of nerve cell in the eye that acts as a receptor of light. Cones are specialized for daylight and color vision. See ROD.

Corpus callosum A very large tract of nerve fibers (axons) connecting the two halves of the cerebral cortex.

Correlated firing The firing of two neurons is said to be correlated when the spikes of one occur more often than by chance at the same time as (or displaced by a constant time from) the spikes of the other. For example, if two neurons always fire at exactly the same time their firing is considered highly correlated.

Dendrite A treelike part of a nerve cell. In most cases dendrites receive signals from other nerve cells. See also AXON.

Disparity The difference in the positions to which a point in space projects on each eye. In the visual system, nerve cells that receive inputs from both eyes often respond to small differences (disparities) between the two inputs. This property makes stereo vision possible.

Dopamine A small chemical molecule that can act as a neurotransmitter.

Dualism The idea that the mind and the brain are separate entities, the mind being in some sense immaterial and obeying laws unknown to the rest of science. The common belief of mankind; probably erroneous.

Electrode An electrical conductor through which electricity enters or leaves some other medium, such as a conducting fluid. In neuroscience, usually a thin conductor (insulated except at its tip) placed near (or occasionally in) a nerve cell. It can be used to pick up signals generated by the cell's electrical activity or to stimulate it, or both.

Electroencephalograph (EEG) The recording of brain waves, the broad electrical activity of the brain, by means of electrodes attached to the skull. Its time resolution is good but its spatial resolution is poor.

Electron microscope A microscope using electrons rather than light. It has a much higher magnification than even a high-powered light microscope. In neuroscience it is commonly used on an extremely thin section of tissue that has been both "fixed" (chemically treated) and dried.

Emergent A system has emergent properties if they are not possessed by its parts. In science, "emergent" does not have mystical overtones. See page 11.

Enzyme A biological catalyst. (A catalyst accelerates a chemical reaction but is unchanged at the end of it.) In almost all cases enzymes are fairly large protein molecules, although some have smaller organic molecules attached to them.

Epigenetic An epigenetic process is one that occurs in the early development of an organism under the influence of its genes.

Event-related potential Potential means voltage. An event-related potential is a changing voltage produced in the brain by an "event" such as a sensory input. It is usually recorded using electrodes on the skull. (See ELECTROENCEPHALOGRAPH.) The signal-to-noise ratio is usually rather poor.

Feature detector A "feature" is the type of stimulus to which a particular neuron in the brain responds, the neuron being loosely described as a feature detector. See page 146.

Filling-in An action by the brain that "guesses" the nature of absent information by assuming that it resembles related information. See BLIND SPOT.

Fovea A depression near the center of the retina where the photoreceptors are very close together, allowing very sharp vision in that region.

Free Will The feeling that one is free to make personal choices.

Functionalist Someone who believes that the best approach to understanding the mind is to study and theorize how it behaves without worrying how its neural components are connected or how they behave, a view often held aggressively by theorists who dislike neuroscience.

GABA A small chemical whose proper name is gamma-amino-butyric acid. It is the major inhibitory neurotransmitter in the forebrain.

GABAergic Using GABA as a neurotransmitter.

Ganglion cell A nerve cell in the retina that receives signals from other neurons in the retina and sends signals to the brain.

Gestalt An organized whole in which the individual parts interact to produce the behavior of the whole. Used in psychology. See pages 36–41.

Glial cell A cell in the nervous system that is not a nerve cell but performs some supporting function. There are several distinct types of glial cells.

Global workspace A term used by Bernard Baars to denote hypothetical brain regions that act as a central information exchange between specialized processors.

Glutamate A small organic chemical (also called "glutamic acid" at low acidity). The major excitatory neurotransmitter in the forebrain.

Gyrus The crest of the bulges on the surface of the folded cerebral cortex. Each particular gyrus has been given a name, such as the angular gyrus.

Hebb's rule Named after Canadian psychologist Donald Hebb. A type of alteration to the strength of a synapse that depends on both the presynaptic activity coming into the synapse and the activity (of some sort) of the receiving neuron on the postsynaptic side. It is important because the alteration to the synapse requires the association in time of two distinct forms of neural activity. See pages 101 and 183.

Hertz A measure of frequency, often abbreviated Hz. One Hertz is one cycle, or event, per second. In the United States, alternating electric current has a frequency of 60 Hertz. Middle C is about 260 Hertz.

Hippocampus A part of the brain, so-called because its shape resembles the tiny aquatic creature known as a sea horse. Sometimes called the ARCHICORTEX. Because of its relatively simple structure it is much studied. It is probably involved in the temporary storage or coding of long-term episodic memory. See pages 83–173.

Homunculus An imaginary person within the brain who perceives objects and events and makes decisions.

Hopfield network A type of simple neural network—named after John Hopfield, who invented it—that feeds back on itself and has symmetrical connections. Because of its symmetry and the way it is adjusted it has an "energy function" associated with it. See pages 182–185.

Hypothalamus A small region in the brain about the size of a pea. It secretes hormones locally and is involved with the control of hunger, thirst, sex, and so on.

Inferotemporal cortex A gyrus in the lower part of the temporal lobe of the cerebral cortex. In the macaque monkey the neurons in this area respond to various complex visual patterns.

Intertectal commissure Also called the "posterior commissure." It includes nerve fibers (axons) that link the superior colliculus on one side to that on the other.

Intralaminar nucleus Any one of a small set of thalamic areas that project mainly to the corpus striatum and also, somewhat diffusely, to many areas of the cerebral cortex.

Ion An atom or small molecule that carries an electric charge. The movement of ions across cell membranes is at the root of the main methods of electrical signalling in the brain.

Kanizsa triangle An optical illusion invented by the Italian psychologist Gaetano Kanizsa. See Figure 2.

Lateral Geniculate Nucleus (LGN) A small part of the thalamus. It relays signals from the eye to the visual cortex. It also receives extensive signals back from the cortex whose exact functions are not yet known.

Lipid A general descriptive term for certain organic molecules that have one water-loving end and one fat-loving end. A double layer of lipid makes up the lipid bilayer that forms the basis of most biological membranes, such as that which surrounds each cell.

Locus ceruleus A pigmented region of the pons (in the brain stem). One of its axons can have an enormous number of synapses, often extending over a large region of the cerebral cortex. Its exact function is not known. Much of it is silent in REM sleep.

Magnetic Resonance Imaging (MRI) A modern noninvasive method of scanning the body (and the brain in particular) that depends on the magnetic resonance of certain atomic nuclei. The standard method produces static two-dimensional maps (often combined to produce a three-dimensional image) with surprisingly good spatial resolution. See page 115. Recent new methods of using the technique can show certain activities in the brain.

Magnocellular (M cells) Having large nerve cells. Originally used in the visual system to describe two of the six layers of the LGN (see also PARVOCELLULAR). The term *M cells* is now used as a general term for primate nerve cells in the retina and the visual cortex whose responses to visual signals are somewhat similar. See pages 125 and 129.

Masking In visual psychology, the effect on the visibility of a brief visual signal of another, usually similar, signal present at about the same time and place.

Microelectrode A very small ELECTRODE used to pick up electrical signals mainly from a single nerve cell.

Micron A unit of length: a thousandth of a millimeter, and so a millionth of a meter. Often written μm (pronounced "mew"). A convenient length, since the wavelengths of visible light are near half a micron, while most cells have a diameter of about 10–50 microns.

Middle temple (MT) A particular visual cortical area in the monkey's brain. Sometimes called "V5" (for the fifth visual area). Its neurons are especially interested in movement.

Molecular biology The study of biology at the molecular level, especially the study of the structure, synthesis, and behavior of proteins and nucleic acids. Now the dominant approach to many biological problems due to its precision and its enormously powerful experimental methods.

Motor cortex The parts of the cerebral cortex concerned mainly with planning and executing movements.

Necker cube A skeletal drawing of a cube that can be perceived in at least two distinct ways. See Figure 4.

Neglect A person suffering from visual neglect (usually because of brain damage) can see in both halves of the visual field but tends to ignore objects on one side if there is something of interest on the other.

Neocortex The "new" cortex. The main part of the cerebral cortex of mammals. Other parts are the paleocortex and archicortex. Usually, when one says "cortex," one means the cerebral neocortex.

NETtalk A neural net designed to learn from examples how written English is pronounced. Discussed at length in Chapter 13.

Neural correlate The neural correlate of some sensation, thought, or action is the nature and behavior of the nerve cells whose activity is closely related to that mental activity. The neural correlates of consciousness have yet to be discovered.

Neural network A computational device made of units that are like very oversimplified neurons. Such units can be connected together in many different ways. The strength of the connections can be altered in an attempt to make the network behave in the desired way. See Chapter 13.

Neuroanatomy The study of the structure of the nervous system and in particular of its neurons and the way they are connected together. A branch of NEUROBIOLOGY.

Neurobiology The biology of the nervous system of animals. By a curious historical accident psychology is not usually considered part of neurobiology. It tends to be taught not in biology departments but in separate academic departments. The number of neurobiologists has grown enormously in the last twenty-five years or so.

Neuron The scientific name for a nerve cell. See Chapter 8.

Neurophysiology A branch of neuroscience dealing with the behavior of the nervous system and its components, in particular how, why, and when nerve cells fire.

NMDA A small chemical—N-methyl-D-aspartate—related to GLUTAMATE. An NMDA receptor is a form of glutamate receptor that also responds to NMDA. It is important in some forms of synaptic modification. See page 101.

Norepinephrine Also called "noradrenaline." A hormone and a neurotransmitter used, for example, by the LOCUS CERULEUS.

Ocular dominance The degree to which a nerve cell in the visual system responds primarily to one eye or the other. (Some nerve cells respond only to the left eye, some only to the right eye, and others respond, to differing extents, to both eyes.)

Oscillations (gamma, 40 Hertz) Neurons, and especially BRAIN WAVES, show somewhat irregular periodicities over a variety of frequency ranges. Those near 10 Hertz are called alpha-rhythms; those near 20 Hertz are sometimes known as beta waves. Those in the 35- 75-Hertz range are sometimes called "gamma oscillations" and sometimes, less accurately, "40-Hertz oscillations." See page 244.

Pacman A solidly colored circular disk with a segment missing. See Figure 2.

Paleocortex The older part of the CEREBRAL CORTEX, largely associated with smell.

Parvocellular (P cells) Having small nerve cells. Originally used in the visual system to describe four of the six layers of the LGN (see also MAGNOCELLULAR). *P cells* is now used as a general term for primate nerve cells in the retina or the visual cortex whose responses to visual signals are somewhat similar. See pages 125 and 129.

Patch-clamping A method for studying the behavior of individual ion channels in a minute area of membrane. See page 118.

PDP Short for Parallel Distributed Processing, a computational technique rather different from that used in the usual type of computer, see Chapter 13. Also used as the name of the group of people (mainly at San Diego) who helped develop such a style of computation.

Perceptron A very simple neural network studied especially by Frank Rosenblatt. See Chapter 13.

PET scan PET stands for Positron Emission Tomography. A technique for studying activity in the living brain by the use of radioactive substances that emit a positron. The result is a rather coarse brain map that shows where the activity (related to some task) is located.

Photon A particle of light. Light (like matter) has both particle-like and wavelike properties.

Photoreceptor A specialized nerve cell that responds to light of a certain range of wavelengths.

Pop-out Pop-out occurs when something in the visual field strikes you almost immediately, independent of the number of distracting objects in the field. See Figure 20.

Positron An elementary particle similar to an electron but with a positive instead of a negative charge. If a positron meets an electron they annihilate each other, producing a pair of gamma-rays (X-rays having a very short wavelength) in the process. Used in PET SCANS.

Potential In neuroscience, this term often means voltage. It is usually measured in millivolts, where 1 millivolt is one thousandth of a volt.

Projects In neuroscience, a nerve cell "projects" to a particular place if its axon ends there. If region A projects to region B this implies that A and B are connected so that neural signals travel *from* A *to* B.

Prosopagnosia The inability to recognize faces, or certain aspects of them, usually due to brain damage.

Protein A large family of biological molecules constructed by stringing amino acids together to form long chains. Proteins, of which there are very many distinct types, are the machine tools of the cell. Enzymes and ion-channels are made of protein, as are other important large biological molecules.

Psychology "The systematic study of behaviour and the mind in man and animals, a discipline which has as yet little coherence. It has many different branches, of which some provide explanations little, if at all, advanced from common sense; others put forward reasonably rigorous scientific theories. Almost all branches are united by their faith in the value of experiments, regardless of the importance or replicability of the results." (Quoted, with permission, from the *International Dictionary of Psychology* by Stuart Sutherland [Continuum Publishing Company, New York, 1989].)

Pulvinar A large part of the thalamus; in primates mainly concerned with vision. Distinct from the LGN, the other visual part of the thalamus.

Pyramidal cell The major type of large nerve cell found in the CEREBRAL CORTEX. It has an apical dendrite that is usually fairly large. Its dendrites have many spines. Its axons form type 1 (excitatory) synapses.

Qualia A philosophical term, the plural of quale. The subjective quality of mental experience, such as the redness of red or the painfulness of pain.

Quantum Mechanics The form of mechanics, invented in the 1920s, that accurately describes the behavior of matter and light, especially that of photons and electrons. Its basic ideas do not conform to everyday common sense. For large objects it is, in most cases, approximated sufficiently well by Newtonian mechanics.

Receptive field The part of the visual field in which a stimulus of the right type can excite a nerve cell in the visual system.

Reductionism The idea that it is possible, at least in principle, to explain a phenomenon in terms of less complicated constituents. The main method of explanation used by the exact sciences. Many people, including some philosophers, dislike it, usually for inadequate reasons.

REM sleep REM stands for rapid eye movement. The other main phases of sleep are grouped under the term *slow wave sleep* or sometimes simply *non-REM* sleep. Hallucinoid dreams are common in REM sleep.

Reticular formation An old-fashioned term for many of the groups of nerve cells in part of the brain stem, especially those concerned with sleep and arousal, and various bodily functions.

Retina A multilayered sheet of nerve cells at the back of each eye. Curiously enough, the photoreceptors are situated in the innermost layer while the ganglion cells, whose axons project to the brain, lie in the outermost layer, nearer to the lens of the eye. For this reason there has to be a gap in the retina through which the axons of ganglion cells can pass on their way to the brain. This gap in the photorecepter layer produces a BLIND SPOT in each eye.

Retinal ganglia cell See GANGLION CELL.

Retinotopic A retinotopic mapping in some region means that neighboring points in the retina are represented by neighboring points in that region. This mapping may be distorted in various ways. Retinotopic mapping is common in those levels of the visual system connected more directly to the eyes.

Rod A type of photoreceptor in the retina that functions mainly in dim light. Rods are of only one type, so that in dim light one cannot perceive color. Rods are absent from the fovea but abundant in the periphery of the retina (see CONE).

Saccade (Sometimes pronounced in the French manner "sack-ard.") A flick of the eyes, resulting in a new fixation point. Saccades are rapid, but they cannot be made more often than about five times a second. Most people move their eyes more frequently than they realize, usually three or four times a second.

Salient An object is said to be salient if it attracts attention; stands out conspicuously.

Scotoma In the visual system, a patch of blindness usually produced by damage to the retina or to part of the visual cortex.

Second messenger Some receptor molecules do not react to the relevant neurotransmitter by opening an ion channel but by producing a biochemical change on the inner side of the cell's membrane that sends a diffusable molecule as a signal to other parts of the cell. Such a signal is called a "second messenger." The process is rather slower than the comparatively fast reactions of most ion channels.

Serial search To be contrasted with POP-OUT. A visual process in which one item (or block of items) in a larger group is attended to one after another, rather than all together.

Serotonin A small organic molecule (also known as 5-HT or 5-hydroxytryptamine) used as a neurotransmitter. It is present in the raphe nucleus in the brain stem that sends axons all over the brain. It may be involved in several types of mental illness.

Signal-to-noise ratio The ratio of the "signal" (the desired information) to the "noise" (the unwanted background information). In a crowded cocktail party the signal-to-noise ratio of the information in one person's conversation is usually rather low.

Slow-wave sleep The relatively dreamless forms of sleep associated with slow waves in the EEG. Sometimes called "non-REM sleep" (see REM SLEEP). The two main phases of sleep (slow wave and REM) alternate during sleep, usually with a ninety-minute cycle time. Slow-wave sleep normally occurs before a period of REM sleep.

SOA Stands for Stimulus Onset Asynchrony. The time between the onset of one stimulus and the onset of another.

Soma The scientific word for "cell body."

Somatosensory Related to information about the parts of the body, both external and internal, and dealing with touch, the sensations of hot and cold, and so on.

Spatial frequency In the visual system, the spacing of a regular grating expressed, for example, as cycles per degree of visual angle. A high spatial frequency implies a finely spaced grating.

Spike An informal term for the short pulse of activity that travels down an axon. Also called an ACTION POTENTIAL.

Spine A word with two distinct meanings, the most common is the vertebrate backbone. A "dendritic spine" is a very small twiglike projection from a dendrite, on which an excitatory synapse is located (see page 99). A typical pyramidal cell has many thousands of them on its dendrites.

Squid In this context, an abbreviation for superconducting quantum interference device, an instrument for detecting changes in the very small magnetic fields produced by the brain. See Chapter 9.

Stellate cell A neuron with a somewhat star-shaped dendritic tree. In the cortex, the type called "spiny stellates" produces excitation. Several other nonspiny types produce inhibition.

Striate cortex So called because it has striations, due to many myelinated axons running roughly parallel to the cortical sheet. Also called area 17 or V1, the first visual area.

Sulcus A groove in the folds of the cortex. Most sulci are given individual names, such as the superior temporal sulcus (also referred to as the STS).

Superior colliculus One of a pair of groups of nerve cells, one on each side, at the top of the brain stem. (The equivalent organ in a lower vertebrate is often called the "tectum.") Superior colliculi are part of the visual system, receiving projections from certain ganglion cells in the eye. Their main function in primates appears to be related to eye movements, but as some of their nerve cells also project to the pulvinar they are probably involved in other forms of visual attention as well.

Synapse The connection between one nerve cell and another. Most of these have a minute gap (between the terminal of the incoming axon and the recipient neuron) across which neurotransmitter molecules can diffuse. See pages 97–100. In some parts of the brain, the dendrites of one cell can form a synapse on the dendrites of another, but such synapses are rare or absent in the CEREBRAL CORTEX.

Thalamus An important region of the forebrain intimately related to the CEREBRAL CORTEX. It has many distinct parts. The main visual regions in primates are the LGN and the PULVINAR. The thalamus is the gateway to the cortex, since all the senses (excepting smell) must relay through it to get to the cortex.

Unconscious inference An expression used by Helmholtz in the nineteenth century implying that the unconscious processes in perception are similar to conscious inference. While this appears broadly true, the neural mechanisms involved are probably quite distinct.

Veridical Meaning, broadly, "as it really is," as inferred by other sources of information, such as that from touching visible objects.

V1, V2, V1 means the first visual area in the cortex, V2 the second, and so on. The nomenclature is somewhat arbitrary, especially as V5 is usually called "MT." So far there is no V6; other abbreviations are used for the many other visual areas of the cortex.

Wernicke's area A region torward the back of the brain of the dominant cerebral hemisphere (the one concerned with language). Part of the human language system. Damage to it produces a characteristic form of aphasia. It is unlikely to be a simple cortical area in the modern sense of the term.

Further Reading

"Of making many books there is no end;
and much study is weariness of the flesh."

—Ecclesiastes 12:12

This is a rather personal selection of books that cover a variety of subjects. Some of the books are suitable for the general reader; others are more difficult. I have grouped them under six broad headings to make it easier for you to select which particular topic you want to follow up. The groupings and subgroupings are necessarily somewhat arbitrary. I have provided short comments on each book to convey something of their character.

GENERAL

Blakemore, Colin. *The Mind Machine*. BBC Books, 1988.

> Blakemore is a British physiologist with a wide interest in both the brain and the mind. This is the book of a BBC television series. It covers many aspects of the mind—even consciousness is mentioned briefly. Very readable.

Changeux, Jean-Pierre. *Neuronal Man: The Biology of Mind*. L. Garey, trans. Pantheon, 1985.

> Changeux is a French molecular biologist with a special interest in neurobiology. His book is a very readable account of many aspects of the brains of men and other animals, with many interesting historical asides. It has rather little to say about consciousness.

Kosslyn, Stephen M., and Olivier Koenig. *Wet Mind: The New Cognitive Neuroscience*. The Free Press, 1992.

> Aimed at the general reader, this book covers many aspects of brain function, such as reading, language, and the control of movement, as well as visual perception. The title is derived from the idea that the mind is what the brain does, the brain being "wet," as opposed to computers, which are "dry." There is something about neural networks but rather little about neurons themselves. Consciousness is discussed in the last chapter. Worth reading, although I have

281

my doubts about Kosslyn's theory of consciousness. The book is clearly written and not too difficult to understand.

Edelman, Gerald M. *Bright Air, Brilliant Fire*. Basic Books, 1992.

Edelman is a molecular biologist who now works on developmental biology and theoretical models of the brain. It aims at a more general audience than his three previous books while covering much the same ground. Some of Edelman's stories, while very familiar to his friends, do not sit so well on the printed page.

THE MIND-BODY PROBLEM

Searle, John R. *The Rediscovery of the Mind*. Bradford Books, MIT Press, 1992.

Searle is a philosopher. His book is about the mind-body problem, but he has detailed objections to the usual A.I. approach. He is not a dualist, preferring to believe that conscious states are simply high-level features of the brain. Searle does not address the problem of how neurons might do this, or how they could encode meaning. I agree with him that we are probably anthropomorphizing the brain and that much of our current ideas about it will not survive a detailed understanding of how it works.

Lockwood, Michael. *Mind, Brain and Quantum: The Compound "I."* Blackwell Pubs., 1989.

Lockwood is an Oxford philosopher. He recognizes that consciousness poses a problem for a materialist but hopes that a proper understanding of the paradoxes of quantum mechanics will help to solve it. He is rather vague on how this might be done, falling back on ideas of Herbert Fröhlich that few scientists find credible. He believes that finding out more about the brain will not help at all. Not easy reading.

Churchland, Paul M. *Matter and Consciousness*. Bradford Books, MIT Press, 1984.

Paul Churchland is a Canadian philosopher now working at San Diego. As he explains, he is an eliminative materialist. He knows more about the brain than most philosophers. I agree with him that many of our present psychological ideas are likely to prove only crude approximations to the truth. Easy to read.

Churchland, Paul M. *A Neurocomputational Perspective: The Nature of Mind and the Structure of Science*. Bradford Books, MIT Press, 1989.

Mainly a series of closely argued essays updating and expanding the author's views on qualia, folk psychology, neural networks, and other topics. The book reflects some of the current disagreements among philosophers about those matters.

Dennett, Daniel C. *Consciousness Explained*. Little, Brown, 1991.

Dennett is a philosopher who knows some psychology and also a little about the brain. He has interesting ideas but appears to be overpersuaded by his own eloquence. His main target is the "Cartesian Theater," the idea that there is one single place in the brain where consciousness resides. In this he is probably right, although it is possible that there are distributed Cartesian theaters. He believes that consciousness is an ongoing process, which he describes as a "multiple drafts" model. This idea has much truth in it. He thinks that it is impossible to distinguish between false accounts by the brain of events that never happened from brain events that did happen but were then falsified. I believe that we could probably make this distinction if we knew exactly what went on in the brain during the process. He is probably unsound about filling-in (see the discussion in my Chapters 4 and 11) and unhelpful about qualia.

Dennett does suggest, in a halfhearted way, a few experiments that might be done to support his ideas. Characteristically, they are all psychological; one would never gather from his book that experimental confirmation by the methods of neuroscience is essential.

Churchland, Patricia Smith. *Neurophilosophy: Toward a Unified Science of the Mind-Brain.* Bradford Books, MIT Press, 1986.

Patricia Churchland is a philosopher. She is one of the first neurophilosophers, which means that she has a detailed knowledge of neurons and the brain, and also of neural networks. The first third of her book is an introduction to neuroscience. The second covers recent developments in the philosophy of science. The third part gives some theories of brain function, now somewhat dated. Written in a breezy, readable style.

Jackendoff, Ray. *Consciousness and the Computational Mind.* Bradford Books, MIT Press, 1987.

Jackendoff is a cognitive scientist with a special interest in language and music. In this book he puts forward his intermediate-level theory of consciousness. This implies, for example, that we are not directly aware of our thoughts but only of the silent speech and imagery that such thoughts produce. See the discussion in my Chapters 2 and 14. Clearly written but not all that easy to grasp on a first reading.

Baars, Bernard. A *Cognitive Theory of Consciousness.* Cambridge University Press, 1988.

Baars is one of the few cognitive scientists who take the problem of consciousness seriously. His book describes a general theory of consciousness—the global workspace—and also gives a wide overview of many aspects of consciousness. Although Baars has some interest in neurons, there is only a little about them in this book. See the discussion of his ideas in my Chapters 2 and 17.

Penrose, Roger. *The Emperor's New Mind.* Oxford University Press, 1989.

Penrose is a distinguished mathematician and theoretical physicist. He believes that the brain can execute processes that no possible Turing-type computer could carry out. He considers physics incomplete because there is as yet no theory of quantum gravity. Penrose hopes that an adequate theory of quantum gravity might explain the mystery of consciousness, but he is uncharacteristically vague as to how it might do so. At bottom his argument is that quantum gravity is mysterious and consciousness is mysterious and wouldn't it be wonderful if one explained the other. Much of the book deals with such topics as Turing machines, Gödel's theorem, quantum theory, and the arrow of time, all explained with great thoroughness and clarity. There is a little about some of the properties of the brain, but practically nothing about psychology. Penrose is a Platonist, a point of view not to everybody's taste. It will be remarkable if his main idea turns out to be true.

Popper, Karl R., and John C. Eccles. *The Self and Its Brain*. Springer-Verlag, 1985.

Popper is a philosopher. Eccles is a neuroscientist. The book is in three parts: the first by Popper, the second by Eccles, and the third a dialogue between the two. Both of them are dualists—they believe in the ghost in the machine. I myself have little sympathy with either of their points of view. They would probably say the same of mine.

Eccles, John C. *Evolution of the Brain: Creation of the Self*. Routledge, 1989.

Mainly about the evolution of the human brain. The later chapters present a more up-to-date account of the author's ideas than that in Popper and Eccles's *The Self and Its Brain*.

Edelman, Gerald M. *The Remembered Present: A Biological Theory of Consciousness*. Basic Books, 1989.

This is the third in a series of academic books that set out the author's ideas. Edelman has a wide knowledge of many aspects of these subjects. He is rather overfond of his own concepts (such as the theory of neuronal group selection, neural Darwinism, and reentrant loops). An enthusiast, noted more for his exuberance than for his clarity.

Humphrey, Nicholas. *A History of the Mind: Evolution and the Birth of Consciousness*. Simon & Schuster, 1992.

Humphrey is a neuroscientist at Cambridge University. His book is easy to read and full of British charm. It leads up to a discussion on consciousness. He stresses the importance of feedback loops (as Edelman does) but is a bit vague as to exactly which ones are crucial for consciousness. He does not consider the idea that neural networks can learn to recognize correlations in their inputs.

Marcel, A. J., and E. Bisiach (eds.). *Consciousness in Contemporary Science*. Oxford University Press, 1988.

A very mixed bag but fairly representative of the variety of ways with which people approach the problem of consciousness. Aimed at an academic audience.

Griffin, Donald R. *Animal Minds*. University of Chicago Press, 1992.

Griffin is a biologist. Are animals conscious? This book is a thoughtful discussion of the problem. Griffin makes a plausible case that at least some of them are, and warns against dogmatic pronouncements on the subject. The question is unlikely to be answered decisively until the neural basis of consciousness is understood.

A.I. AND NEURAL NETWORKS

Boden, Margaret A. *Artificial Intelligence and Natural Man.* Basic Books, 1977.

Boden is both a philosopher and a psychologist. Her book is a good description of A.I. as it was at that time. It also discusses its wider implications.

Winston, Patrick Henry. *Artificial Intelligence,* 3rd ed. Addison-Wesley Publishing Company, 1992.

A useful textbook on the subject.

Minsky, Marvin. *The Society of Mind.* Simon & Schuster, 1985.

Minsky is one of the fathers of Artificial Intelligence. This rather rambling book sets out his mature thoughts on how the mind works. It reads rather as though Minsky was thinking aloud. His title encapsulates his basic idea but there are also many seminal remarks, on a variety of topics, scattered throughout the book. Practically nothing about the brain.

Newell, Allen. *Unified Theories of Cognition.* Harvard University Press, 1990.

Newell believed in the possibility of a general theory of cognition, which many would regard as an unlikely one. With his colleagues he devised an architecture for general human cognition they called SOAR. It is constrained by general considerations about the brain—such as the times neurons take to act—but otherwise has little relationship to neuroscience. SOAR deals mainly with thinking, intelligence, and immediate behavior, but not with perception. Newell claims that it provides a theory of "awareness" but not of "consciousness," by which he means qualia. SOAR is mostly concerned with processes that take about a second or more, whereas I have concentrated on those that take less time than this. SOAR has more brainlike properties than many general models based on A.I. Whether it will turn out to have any resemblance to the way the brain really acts remains to be seen.

Blake, Andrew, and Tom Truscianko (eds.). *A.I. and the Eye.* New York: Wiley, 1990.

This is a collection of papers delivered at an international conference that brought together visual psychologists and workers in Artificial Intelligence. The papers show little evidence that the A.I. approach will do much to help us understand the brain, but they do illustrate some of the attempts being made.

Allman, William F. *Apprentices of Wonder: Inside the Neural Network Revolution.* Bantam Books, 1989.

A breezy book by a science journalist, with chitchat about the people involved. An easy way to learn a little more about neural networks and how they originated.

Caudill, Maureen, and Charles Butler. *Naturally Intelligent Systems.* Bradford Books, MIT Press, 1990.

Neural networks in a nutshell. Very clearly written by two "networkers," this is a good, fairly simple introduction for those who want to learn more about them. The book has a useful glossary. Their term *neurodes,* for the unit of neural nets, is not widely used.

Bechtel, William, and Adele Abrahamsen. *Connectionism and the Mind: An Introduction to Parallel Processing in Networks.* Basil Blackwell, 1991.

A fairly readable introduction for students. Although mainly about networks, the book includes some discussion of more general issues. Very little about neurons; nothing about consciousness.

Churchland, Patricia S., and Terrence J. Sejnowski. *The Computational Brain: Models and Methods on the Frontiers of Computational Neuroscience.* Bradford Books, MIT Press, 1992.

Written by two of my close associates at San Diego, this book not only describes modern ideas about computation and neural networks but also discusses examples of how they can be applied to real biological systems. Essential reading for anyone wishing to follow up the very simple introduction to neural networks given in my Chapter 13.

Zornetzer, Steven F., Joel L. Davis, and Clifford Lau (eds.). *An Introduction to Neural and Electronic Networks.* Academic Press, 1990.

This book covers real neurons, silicon neurons, and neural network models. The various articles are written by many of the leading workers in the different fields, from molecules to mathematics. Gives a good idea of the range of approaches now being studied. Not for beginners.

Rumelhart, David E., James L. McClelland, and the PDP Research Group. *Parallel Distributed Processing*, vols. 1 and 2. Bradford Books, MIT Press, 1986.

The book that launched the neural network revolution, becoming an academic best-seller in the process. Now somewhat dated. The four introductory chapters and the final chapter give good overviews of the state of the subject at that time.

Abeles, M. *Corticonics: Neural Circuits of the Cerebral Cortex.* Cambridge University Press, 1991.

The author is an Israeli neurophysiologist. The title has been concocted from "cortex" and "electronics." The book presents a number of interesting arguments about possible general properties of the neural circuits of the brain. Not for beginners.

Schwartz, Eric L. (ed.). *Computational Neuroscience.* Bradford Books, MIT Press, 1990.

A multiauthor work, ranging from biological systems to artificial neural networks. This academic book illustrates rather well the ferment of activity produced by the neural network revolution.

COGNITIVE SCIENCE

Gardner, Howard. *The Mind's New Science: A History of the Cognitive Revolution.* Basic Books, 1985.

A broad description of cognitive science and its origins. Not difficult to read.

Johnson-Laird, Philip N. *Mental Models,* Harvard University Press, 1983, and *The Computer and the Mind: An Introduction to Cognitive Science,* Harvard University Press, 1988.

> Johnson-Laird is a British cognitive psychologist, now at Princeton. *Mental Models* is mainly about language and inference, with a short section on consciousness and computation. *The Computer and the Mind* deals with a wider variety of topics, including visual perception. Both books are thoughtful yet fairly easy to read. See my remarks about his ideas in Chapters 2 and 14.

Posner, Michael I. (ed.). *Foundations of Cognitive Science.* Bradford Books, MIT Press, 1989.

> An academic book for those who want to know what cognitive science is all about. Nothing about consciousness. Only one chapter mentions neurons.

Sutherland, Stuart. *The International Dictionary of Psychology.* Macmillan Ltd., 1989.

> Covers most of the technical terms used in psychology and some of the subjects closely related to it. Sutherland has strong opinions about certain branches, such as psychoanalysis. His definition of love is unconventional.

Hebb, D. O. *Organization of Behavior.* (First published 1949.) Wiley, 1964.

> Remembered mainly for its clear statement of what is now called "Hebb's rule" (see Chapter 13) and for his rather less clear proposal of reverberatory circuits.

James, William. *The Principles of Psychology.* (First published 1890.) Harvard University Press, 1981.

> Undoubtedly a classic. Still worth reading in spite of its age. Shows that consciousness was an important topic in psychology in those days. Parts of it are quoted in my Chapter 2.

VISUAL PERCEPTION

Rock, Irvin. *Perception.* Scientific American Library, distributed by W. H. Freeman, 1984.

> An excellent introduction to visual perception. The author is a psychologist well known for his research on the behavior of our visual system. A thoughtful book yet easy to read and well illustrated. No mention of consciousness and only a little about neurons and the brain.

Sekuler, Robert, and Randolph Blake. *Perception,* 3rd ed. McGraw-Hill, 1993.

> Both authors are psychologists. Their book deals with all the senses but mostly with vision. Aimed at students but fairly easy for a layman to understand. Mainly psychology, with a little about the brain.

Marr, David. *Vision.* W. H. Freeman, 1983.

> Destined to become a classic, mainly because of the clarity of the author's thinking and the forceful way he puts forward his point of view. Both his general attitude and many of his detailed suggestions now appear somewhat dated.

In spite of this, his insistence on a careful analysis of the problem and on the importance of producing an explicit model is likely to remain. It was published posthumously.

Kanizsa, Gaetano. *Organization in Vision: Essays on Gestalt Perception*. Praeger, 1979.

Kanizsa was an Italian psychologist. The book has many striking figures, mainly devised by the author, that illustrate different aspects of the behavior of our visual system. It could well become a classic.

Petry, Susan, and Glenn E. Meyer (eds.). *The Perception of Illusory Contours*. Springer-Verlag, 1987.

A multiauthored work based on a conference. Displays many examples of illusory contours and almost as many ideas about them. Only for those keenly interested in the topic.

Johnson, Mark H., and John Morton. *Biology and Cognitive Development: The Case of Face Recognition*. Blackwell, 1991.

A well-written book on a topic of interest to almost everyone. Scholarly yet a pleasure to read. Keeps off the difficult topic of consciousness in human babies.

Weiskrantz, L. *Blindsight: A Case Study and Implications*. Oxford University Press, 1986.

An authoritative general overview of the subject, as well as a detailed account of some of the author's previously unpublished work. Useful as a background to more recent developments.

Kosslyn, Stephen Michael. *Ghosts in the Mind's Machine*. W. W. Norton, 1983.

Kosslyn is one of the pioneers of the scientific study of mental imagery, an interesting subject about which I have said practically nothing. Fairly easy to read.

Baddeley, Alan. *Human Memory: Theory and Practice*. Allyn and Bacon, 1990.

Baddeley is a British psychologist. Covers many aspects of memory, often with a historical background. Rather detailed but written in a fairly readable manner. A little about brain damage and neural networks but nothing about real neurons.

Julesz, Bela. *Foundations of Cyclopean Perception*. University of Chicago Press, 1971.

Julesz is a Hungarian psychologist who worked for many years at Bell Telephone Laboratories. His invention of the random dot stereogram revolutionized our ideas about stereo vision. This very detailed account of his research has become a classic.

Gregory, R. L., and E. H. Gombrich (eds.). *Illusion in Nature and Art*. Duckworth, 1973.

Gregory is a British visual psychologist; Gombrich a well-known art critic. Written, with four British co-authors, for the general reader. The book is full of interesting observations, both about nature and about art.

Barlow, Horace, Colin Blakemore, and Miranda Watson-Smith. *Images and Understanding.* Cambridge University Press, 1990.

> I see I wrote the preface, which states that the book is "a feast for everyone." The book covers a great variety of topics, from neurons and brains to moving pictures, dance, and caricatures.

NEUROSCIENCE

Dowling, John E. *The Retina: An Approachable Part of the Brain.* Harvard University Press, 1987.

> Dowling has worked for many years on the retina. His book is a well-written overview of the subject, aimed mainly at students.

Hubel, David H. *Eye, Brain and Vision.* Scientific American Library, distributed by W. H. Freeman, 1987.

> A very readable and well-illustrated account of the early stages of the mammalian visual system by a distinguished neurophysiologist. Hubel is a recent convert to psychology (psychophysics). He is rather reluctant to venture beyond cortical areas V1 and V2. Nothing about consciousness.

Zeki, Semir. *A Vision of the Brain.* Blackwell Scientific Publications, 1993.

> Zeki is a well-known British neuroscientist who pioneered the exploration of parts of the monkey's visual system beyond areas V1 and V2. The book is centered on his own work and especially on his interest in color vision. There is rather little about the inferotemporal cortex. His short chapters combine clear accounts of a great variety of experimental details with many thoughtful general observations. The last chapter is about consciousness in relation to vision. Mainly aimed at students, but also suitable for anyone who wants to know more about the neuroscience of vision. Written in an easy, readable style.

Blakemore, Colin (ed). *Vision: Coding and Efficiency.* Cambridge University Press, 1990.

> The book is a series of scientific papers honoring Horace Barlow, who has put forward many seminal ideas on the visual system. It covers a wide variety of topics related to vision. Aimed at an academic audience. The short piece by Barlow, at the beginning, is a joy to read.

Farah, Martha J. *Visual Agnosia: Disorders of Object Recognition and What They Tell Us about Normal Vision.* Bradford Books, MIT Press, 1990.

> A thoughtful, well-written academic book. Rather too detailed for the general reader but important for students of vision.

Damasio, Hanna, and Antonio R. Damasio. *Lesions Analysis in Neuropsychology.* Oxford University Press, 1989.

> Written by two neurologists, the book outlines how various scanning methods (such as MRI) can tell us about human brains that have suffered damage of one sort or another. It discusses the advantages and limitations of the lesion

method and describes some of the important results it has given. The authors outline their idea of "convergence zones," which appear to be located not only in many cortical areas but also in most of the brain regions associated with the cortex. Many interesting pictures of damaged brains. Mainly for medical workers and scientists.

Dudai, Yadin. *The Neurobiology of Memory: Concepts, Findings, Trends.* Oxford University Press, 1989.

Dudai is a neurobiologist. His book is aimed mainly at an academic audience. It ranges from man to the Californian sea slug. A clearly written, thoughtful book.

Squire, Larry R. *Memory and Brain.* Oxford University Press, 1987.

Squire is a neuropsychologist. Although aimed at scientists and students, it gives a readable outline of what is known about many different aspects of memory.

Dowling, John E. *Neurons and Networks: An Introduction to Neuroscience.* Belknap Press of Harvard University Press, 1992.

This book is not about neural networks as theoretical models of the brain but is rather a general introduction to neuroscience. It is based on an introductory course the author has taught at Harvard, and is aimed at a similar readership.

Shepherd, Gordon M. (ed.). *The Synaptic Organization of the Brain,* 3rd ed. Oxford University Press, 1990.

This is the latest edition of a well-known textbook. A multidisciplinary account of neurons, their components, and their organization into circuits. Covers mostly the better-understood parts of the human brain. Rather too detailed and too difficult for the general reader.

Nicholls, John G., A. Robert Martin, and Bruce G. Wallace. *From Neuron to Brain,* 3rd ed. Sinauer Associates, 1992.

This is the latest edition of a standard textbook. It sets out much basic information about nervous systems. One section describes in some detail the early stages of the mammalian visual system, from the retina (via the LGN) to the visual cortex, but there is practically nothing about the problem of how we see.

Kandel, Eric R., James H. Schwartz, and Thomas M. Jessell (eds.). *Principles of Neural Science,* 3rd ed. Appleton and Lange, 1991.

A standard textbook, aimed at students of biology, behavior, and medicine. The book covers many aspects of the brain, written by a variety of authors. There are several chapters about the visual system. The one by Kandel on visual perception points out that it is a creative process, giving examples from visual psychology. Kandel also discusses the binding problem, attention, the 40-Hertz oscillations and their bearing on visual awareness.

Groves, Philip M., and George V. Rebec. *Introduction to Biological Psychology,* 4th ed. William C. Brown, 1992.

A textbook covering many aspects of the brain, from seeing to sex. Aimed at college students.

Nauta, Walle J. H., and Michael Feirtag. *Fundamental Neuroanatomy*. W. H. Freeman, 1986.

Nauta is a distinguished neuroanatomist, Feirtag a scientific journalist. Aimed at medical students but a useful introduction for neuroscientists. The complexities of the subject make the book rather too difficult for the general reader but its very clear illustrations make it worth consulting.

Peters, Alan, and Edward G. Jones (eds.). *Cerebral Cortex*, vols. 1–9. Plenum, 1984–1991.

The standard reference work, written by many authors. The first volume appeared in 1984, the most recent (vol. 9) in 1991. The earlier parts are now somewhat dated.

Jones, Edward G. *The Thalamus*. Plenum, 1985.

Still the standard work on the thalamus. An updated version would be welcome.

Steriade, Mircea, Edward G. Jones, and Rodolfo R. Llinas (eds.) *Thalamic Oscillations and Signalling*. Wiley, 1990.

A scholarly account of the topic by three well-known authorities. Not easy reading. Written before much of the present interest in 40-Hertz oscillations.

Levitan, Irwin B., and Leonard K. Kaczmarek. *The Neuron: Cell and Molecular Biology*. Oxford University Press, 1991.

Aimed at advanced students, the book has an extended discussion of ion channels. It conveys very well the great molecular complexity of just a single neuron.

Hall, Zach W., et al. *An Introduction to Molecular Neurobiology*. Sinauer Associates, 1992.

A good, rather solid textbook, aimed at an academic audience. Gives a good overview of the many ramifications of the subject and its complexity.

References

PART I
1: Introduction
Pages 3–12

1. Popper, K. R., and Eccles, J. C. (1985). *The Self and Its Brain*. New York: Springer-Verlag.
2. Eccles, J. D. (1986). Do mental events cause neural events analogously to the probability fields of quantum mechanics? *Proc Roy Soc Lond* B 227:411–428.
3. Barlow, H. B. (1972). Single units and sensation: a neuron doctrine for perceptual psychology? *Perception* 1:371–394.
4. Jacob, F. (1977). Evolution and tinkering. *Science* 196:1161–1166.

2: The General Nature of Consciousness
Pages 13–22

1. Kosslyn, S. M. (1983). *Ghosts in the Mind's Machine*. New York: W. W. Norton & Co.
2. Johnson-Laird, P. N. (1983). *Mental Models*. Cambridge, MA: Harvard Univ Press.
3. Johnson-Laird, P. N. (1988). *The Computer and the Mind: An Introduction to Cognitive Science*. Cambridge, MA: Harvard Univ Press.
4. Jackendoff, R. (1987). *Consciousness and the Computational Mind*. Cambridge, MA: Bradford Books, MIT Press.
5. Baars, B. (1988). A *Cognitive Theory of Consciousness*. Cambridge, England: Cambridge Univ Press.
6. Crick, F., and Koch, C. (1990). Towards a neurobiological theory of consciousness. *Seminars Neurosc* 2:263–275.

3: Seeing
Pages 23–33

1. Kanizsa, G. (1979). *Organization in Vision: Essays on Gestalt Perception*. New York: Praeger Publishers.

4: The Psychology of Vision
Pages 35–57

1. Rock, I., and Palmer, S. (1990). The legacy of Gestalt psychology. *Sc Am* Dec:84–90.
2. Nakayama, K., and Shimojo, S. (1992). Experiencing and perceiving visual surfaces. *Science* 257:1357–1363.
3. Chaudhuri, A. (1990). Modulation of the motion aftereffect by selective attention. *Nature* 344:60–62.
4. Ramachandran, V. S. (1992). Blind spots. *Sc Am* 266:86–91.
5. Ramachandran, V. S., and Gregory, R. L. (1991). Perceptual filling in of artificially induced scotomas in human vision. *Nature* 350:699–702.
6. Ramachandran, V. S. (1993). Filling in gaps in perception: Part 2., scotomas and phantom limbs. *Curr Direct Psychol Sc* 2:56–65.

5: Attention and Memory
Pages 59–70

1. Posner, M. I., and Presti, D. E. (1987). Selective attention and cognitive control. *Trends Neurosc* 10:13–17.
2. Luck, S. J., Hillyard, S. A., Mangun, G. R., and Gazzaniga, M. S. (1989). Independent hemispheric attentional systems mediate visual search in split brain patients. *Nature* 342:543–545.
3. Yantis, S. (1992). Multi-element visual tracking: attention and perceptual organization. *Cogn Psychol* 24:295–340.
4. Baylis, G. C., and Driver, J. (1993). Visual attention and objects: evidence for hierarchical coding of location. *J Exp Psychol* 19:1–20.
5. Julesz, B. (1990). Early vision is bottom-up, except for focal attention. *Cold Spring Harbor Symposia on Quantitative Biology, The Brain* 55:973–978.
6. Treisman, A. M., Sykes, M., and Gelade, G. (1977). Selective attention and stimulus integration. In: S. Dornic (ed.), *Attention and Performance VI* (pp. 333–361). Hillsdale, NJ: Lawrence Erlbaum.
7. Egeth, H., Virzi, R. A., and Garbart, H. (1984). Searching for conjunctively defined targets. *J Exp Psychol: Human Perception and Performance* 10:32–39.
8. Treisman, A., and Gormican, S. (1988). Feature analysis in early vision: evidence from search asymmetries. *Psychol Rev* 95:15–48.
9. Treisman, A., and Schmidt, H. (1982). Illusory conjunctions in the perception of objects. *Cogn Psychol* 14:107–141.
10. Cave, K. R., and Wolfe, J. M. (1990). Modeling the role of parallel processing in visual search. *Cogn Psychol* 22:225–271.
11. Duncan, J., and Humphreys, G. W. (1989). Visual search and stimulus similarity. *Psychol Rev* 96:433–458.
12. Dudai, Y. (1989). *The Neurobiology of Memory: Concepts, Findings, Trends.* Oxford, England: Oxford Univ Press.
13. Sperling, G. (1960). The information available in brief visual presentations. *Psychol Monographs* 74:Whole no. 498.
14. Baddeley, A. (1990). *Human Memory: Theory and Practice.* Needham Hgts, MA: Allyn & Bacon, Inc.
15. Shallice, T., and Vallar, G. (1990). The impairment of auditory-verbal short-term

storage. In: G. Vallar and T. Shallice (eds.), *Neuropsychological Impairments of Short-term Memory* (pp. 11–53). Cambridge, England: Cambridge Univ Press.

6: The Perceptual Moment: Theories of Vision
Pages 71–78

1. Libet, B. (1985). Unconscious cerebral initiative and the role of conscious will in voluntary action. *Behav Brain Sc* 8:529–566.
2. Efron, R. (1967). The duration of the present. *Annals NY Acad Sc* 138:367–915.
3. Reynolds, R. I. (1981). Perception of an illusory contour as a function of processing time. *Perception* 10:107–115.
4. Ramachandran, V. S., personal communication. See Ramachandran, V. S. (1990). In: A. Blake and T. Troscianko (eds.), *A. I. and the Eye* (pp. 21–77). Chichester, England: John Wiley & Sons, Inc.

PART II
8: The Neuron
Pages 91–105

1. Hollmann, M., and Heinemann, S. (1993). Cloned glutamate receptors. *Ann Rev Neurosc* 17:31–108

9: Types of Experiment
Pages 107–119

1. Ojemann, G. A. (1990). Organization of language cortex derived from investigations during neurosurgery. *Sem Neurosc* 2:297–305.
2. Crick, F., and Jones, E. (1993). Backwardness of human neuroanatomy. *Nature* 361:109–110.
3. Neville, H. J. (1990). Intermodal competition and compensation in development: evidence from studies of the visual system in congenitally deaf adults. *Ann NY Acad Sci* 608:71–91.
4. Roe, A. W., Pallas, S. L., Kwon, Y. H., and Sur, M. (1992). Visual projections routed to the auditory pathway in ferrets: receptive fields of visual neurons in primary auditory cortex. *J Neurosc* 12:3651–3664.
5. Pardo, J. V., Pardo, P. J., Janer, K. W., and Raichle, M. E. (1990). The anterior cingulate cortex mediates processing selection in the Stroop attentional conflict paradigm. *Proc Natl Acad Sci USA* 87:256–259.
6. Clark, V. P., Courchesne, E., and Grafe, M. (1992). In vivo myeloarchitectonic analysis of human striate and extrastriate cortex using magnetic resonance imaging. *Cerebral Cortex* 2:417–424.
7. Neher, E., and Sakmann, B. (1992). The patch clamp technique. *Sc Am* March:44–51.

10: The Primate Visual System—Initial Stages
Pages 121–138

1. Sparks, D. L., Lee, D., and Rohrer, W. H. (1990). Population coding of the direction, amplitude, and velocity of saccadic eye movements by neurons in the superior colliculus. *Cold Spring Harbor Symposia on Quantitative Biology, The Brain* 55:805–811.
2. Schiller, P. H., and Logothetis, N. K. (1990). The color-opponent and broadband channels of the primate visual system. *Trends Neurosc* 13:392–398.

11: The Visual Cortex of Primates
Pages 139–159

1. LeVay, S., Hubel, D. H., and Wiesel, T. N. (1975). The pattern of ocular dominance columns in macaque visual cortex revealed by a reduced silver stain. *J Comp Neurol* 159:559–575.
2. Mitchison, G. (1991). Neuronal branching patterns and the economy of cortical wiring. *Proc Roy Soc Lond* B 245:151–158.
3. Grosof, D. H., Shapley, R. M., and Hawken, M. J. (1992). Monkey striate responses to anomalous contours? *Investigative Ophthalm Vis Sc* S 33:1257.
4. Von der Heydt, R., Peterhans, E., and Baumgartner, G. (1984). Illusory contours and cortical neuron responses. *Science* 224:1260–1262.
5. Felleman, D. J., and Van Essen, D. C. (1991). Distributed hierarchical processing in the primate cerebral cortex. *Cerebral Cortex* 1:1–47.
6. Allman, J., Miezin, F., and McGuinness, E. (1985). Direction- and velocity-specific responses from beyond the classical receptive field in the middle temporal visual area (MT). *Perception* 14:105–126.
7. Born, R. T., and Tootell, R. B. H. (1992). Segregation of global and local motion processing in primate middle temporal visual area. *Nature* 357:497–499.
8. Adelson, E. H., and Movshon, J. A. (1982). Phenomenal coherence of moving visual patterns. *Nature* 300:523–525.
9. Stoner, G. R., and Albright, T. D. (1992). Neural correlates of perceptual motion coherence. *Nature* 358:412–414.
10. Zeki, S. (1983). Colour coding in the cerebral cortex: the reaction of cells in monkey visual cortex to wavelengths and colours. *Neurosc* 9:741–765.

12: Brain Damage
Pages 161–175

1. Bisiach, E. and Luzzatti, C. (1978). Unilateral neglect, representational schema, and consciousness. *Cortex* 14:129–133.
2. Sacks, O., and Wasserman, R. (1987). The case of the colorblind painter. *NY Rev of Books* 34:25–34.
3. Damasio, A. R., Tranel, D., and Damasio, H. (1990). Face agnosia and the neural substrates of memory. *Annu Rev Neurosci* 13:89–109.
4. Tranel, D., Damasio, A. R., and Damasio, H. (1988). Intact recognition of facial expression, gender, and age in patients with impaired recognition of face identity. *Neurology* 38:690–696.
5. Hess, R. H., Baker, C. L., and Zihl, J. (1989). The "motion-blind" patient: low-level spatial and temporal filters. *J Neurosc* 9:1628–1640.

6. Warrington, E. K., and Taylor, A. M. (1978). Two categorical stages of object recognition. *Perception* 7:695–705.

7. Humphreys, G. W., and Riddock, M. J. (1987). *To See but Not to See: A Case Study of Visual Agnosia*. London: Lawrence Erlbaum Assoc.

8. Brown, J. W. (1983). The microstructure of perception: physiology and patterns of breakdown. *Cogn Brain Theory* 6:145–184.

9. Damasio, A. R., Damasio, H., Tranel, D., and Brandt, J. P. (1990). Neural regionalization of knowledge access: preliminary evidence. *Cold Spring Harbor Symposia on Quantitative Biology, The Brain* 55:1039–1067.

10. Bogen, J. E. (1993). The callosal syndromes. In: K. M. Heilman and E. Valenstein (eds.), *Clinical Neuropsychology*, 3rd ed. (pp. 337–407). Oxford, England: Oxford Univ Press.

11. Sperry, R. W. (1961). Cerebral organization and behaviour. *Science* 133:1749–1757.

12. Weiskrantz, L. (1986). *Blindsight*. Oxford, England: Oxford Univ Press.

13. Stoerig, P., and Cowey, A. (1989). Wavelength sensitivity in blindsight. *Nature* 342:916–918.

14. Fendrich, R., Wessinger, C. M., and Gazzaniga, M. S. (1992). Residual vision in a scotoma: implications for blindsight. *Science* 258:1489–1491.

15. Tranel, D., and Damasio, A. R. (1988). Non-conscious face recognition in patients with face agnosia. *Behav Brain Res* 30:235–249.

13: Neural Networks
Pages 177–199

1. Anderson, C. H., and Van Essen, D. C. (1987). Shifter circuits: a computational strategy for dynamic aspects of visual processing. *Proc Natl Acad Sci USA* 84:6297–6301.

2. Newell, A. (1990). *Unified Theories of Cognition*. Cambridge, MA: Harvard Univ Press.

3. McCulloch, W. S., and Pitts, W. (1943). A logical calculus of the ideas imminent in neural nets. *Bulletin of Mathematical Biophysics* 5:115–137.

4. Rosenblatt, F. (1962). *Principles of Neurodynamics*. New York: Spartan Books.

5. Minsky, M., and Papert, S. (1969). *Perceptrons: An Introduction to Computational Geometry*. Cambridge, MA: MIT Press.

6. Hopfield, J. J. (1982). Neural networks and physical systems with emergent collective computational abilities. *Proc Natl Acad Sci USA* 79:2554–2558.

7. Hebb, D. O. (1964). *Organization of Behavior*. New York, NY: John Wiley & Sons, Inc.

8. Crick, F. H. C., and Mitchison, G. (1983). The function of dream sleep. *Nature* 304:111–114.

9. Crick, F., and Mitchison, G. (1986). REM sleep and neural nets. *J Mind Behav* 7:229–249.

10. Willshaw, D. (1981). Holography, associative memory, and inductive generalization. In: G. E. Hinton and J. A. Anderson (eds.), *Parallel Models of Associative Memory* (pp. 83–104). Hillsdale, NJ: Lawrence Erlbaum Associates.

11. Rumelhart, D. E., McClelland, J. L., and the PDP Research Group (eds.) (1986). *Parallel Distributed Processing*. Cambridge, MA: Bradford Books, MIT Press.

12. Sejnowski, T. J., and Rosenberg, C. R. (1987). Parallel networks that learn to pronounce English text. *Complex Systems* 1:145–168.

13. Lehky, S. R., and Sejnowski, T. J. (1990). Neural network model of visual cortex for determining surface curvature from images of shaded surfaces. *Proc Roy Soc Lond* B 240:251–278.

14. Zipser, D. (1992). Identification models of the nervous system. *Neurosc* 47:853–862.

PART III
15: Some Experiments
Pages 215–230

1. Heywood, C. A., Cowey, A., and Newcombe, F. (1991). Chromatic discrimination in a cortically colour blind observer. *Europ J Neurosc* 3:802–812.

2. Newsome, W. T., Britten, K. H., and Movshon, J. A. (1989). Neuronal correlates of a perceptual decision. *Nature* 341:52–54.

3. Salzman, C. D., Murasugi, C. M., Britten, K. H., and Newsome, W. T. (1992). Microstimulation in visual area MT: effects on direction discrimination performance. *J Neurosci* 12:2331–2355.

4. Duensing, S., and Miller, B. (1979). The Cheshire cat effect. *Perception* 8:269–273.

5. Logothetis, N. K., and Schall, J. D. (1989). Neuronal correlates of subjective visual perception. *Science* 245:761–763.

6. Ramachandran, V. S. (1991). Form, motion, and binocular rivalry. *Science* 251:950–951.

7. Piantanida, T. P. (1985). Temporal modulation sensitivity of the blue mechanism: measurements made with extraretinal chromatic adaptation. *Vis Res* 25:1439–1444.

8. Pritchard, R. M., Heron, W., and Hebb, D. O. (1960). Visual perception approached by the method of stabilized images. *Canad J Psychol* 14:67–77.

9. Fiorani, M., Rosa, M. G. P., Gattass, R., and Rocha-Miranda, C. E. (1992). Dynamic surrounds of receptive fields in primate striate cortex: a physiological basis for perceptual completion? *Proc Natl Acad Sci USA* 89:8547–8551.

10. Livingstone, M. S., and Hubel, D. H. (1981). Effects of sleep and arousal on the processing of visual information in the cat. *Nature* 291:554–561.

11. Moran, J., and Desimone, R. (1985). Selective attention gates visual processing in the extrastriate cortex. *Science* 229:782–784.

12.. Spitzer, H., Desimone, R., and Moran, J. (1988). Increased attention enhances both behavioral and neuronal performance. *Science* 240:338–340.

13. Robinson, D. L., and Petersen, S. E. (1992). The pulvinar and visual salience. *Trends Neurosc* 15:127–132.

14. Anderson, C. H., and Van Essen, D. C. (1987). Shifter circuits: a computational strategy for dynamic aspects of visual processing. *Proc Natl Acad Sci USA* 84:6297–6301.

15. Libet, B., Pearl, D. K., Morledge, D. E., Gleason, C. A., Hosobuchi, Y., and Barbaro, N. M. (1991). Control of the transition from sensory detection to sensory awareness in man by the duration of a thalamic stimulus. *Brain* 114:1731–1757.

16: Mainly Speculation
Pages 231–242

1. Milner, P. M. (1974). A model for visual shape recognition. *Psychol Rev* 6:521–535.
2. Douglas, K. L., and Rockland, K. S. (1992). Extensive visual feedback connections from ventral inferotemporal cortex. In: *Society for Neuroscience Abstr* 169. 10.
3. Edelman, G. M. (1990). *The Remembered Present: A Biological Theory of Consciousness.* New York: Basic Books.
4. Connors, B. W., and Gutnick, M. J. (1990). Intrinsic firing patterns of diverse neocortical neurons. *Trends Neurosc* 13:99–104.
5. Magleby, K. L., and Zengel, J. E. (1982). A quantitative description of stimulation-induced changes in transmitter release at the frog neuromuscular junction. *J Gen Physiol* 80:613–638.
6. Zucker, R. S. (1989). Short-term synaptic plasticity. *Ann Rev Neurosc* 12:13–31.
7. Tömböl, T. (1984). Layer VI cells. In: A. Peters and E. G. Jones (eds.), *Cerebral Cortex,* vol 1: *Cellular Components of the Cerebral Cortex* (pp. 479–519). New York: Plenum Press
8. Goldman-Rakic, P. S., Funahashi, S., and Bruce, C. J. (1990) Neocortical memory circuits. *Cold Spring Harbor Symposia on Quantitative Biology, The Brain* 55:1025–1038.
9. Fuster, J. M. (1989). *The Prefrontal Cortex,* 2nd ed. New York: Raven Press.

17: Oscillations and Processing Units
Pages 243–253

1. Freeman, W. J. (1988). Nonlinear neural dynamics in olfaction as a model for cognition. In E. Basar (ed.), *Dynamics of Sensory and Cognitive Processing by the Brain* (pp. 19–29). Berlin: Springer.
2. Gray, C. M., and Singer, W. (1989). Stimulus-specific neuronal oscillations in orientation columns of cat visual cortex. *Proc Natl Acad Sci USA* 86:1698–1702.
3. Eckhorn, R., Bauer, R., Jordan, W., Brosch, M., Kruse, W., Munk, M., and Reitboeck, H. J. (1988). Coherent oscillations: a mechanism of feature linking in the visual cortex? *Biol Cybern* 60:121–130.
4. Gray, C. M., König, P., Engel, A. K., and Singer, W. (1989). Oscillatory responses in cat visual cortex exhibit inter-columnar synchronization which reflects global stimulus properties. *Nature* 338:334–337.
5. Engel, A. K., Kreiter, A. K., König, P., and Singer, W. (1991). Synchronization of oscillatory neuronal responses between striate and extrastriate visual cortical areas of the cat. *Proc Natl Acad Sci USA* 88:6048–6052.
6. Engel, A. K., König, P., Kreiter, A. K., and Singer, W. (1991). Interhemispheric synchronization of oscillatory neuronal responses in cat visual cortex. *Science* 252:1177–1179.
7. Crick, F., and Koch, C. (1990). Towards a neurobiological theory of consciousness. *Seminars Neurosc* 2:263–275.
8. Livingstone, M. S. (1991). Visually-evoked oscillations in monkey striate cortex. *Soc for Neuroscience Conf Proc.*
9. Kreiter, A. K., and Singer, W. (1992). Oscillatory neuronal responses in the visual cortex of the awake macaque monkey. *Europ J Neuroscience* 4:369–375.

10. Bair, W., Koch, C., Newsome, W., and Britten, K. (1993). Power spectrum analysis of MT neurons in the awake monkey. In: F. Eeckman (eds.), *Computation and Neural Systems 92* (In press). Norwell, MA: Kluwer Academic Publ.

11.. Murthy, V. N., and Fetz, E. E. (1992). Coherent 25- to 35-Hz oscillations in the sensorimotor cortex of awake behaving monkeys. *Proc Natl Acad Sci USA* 89:5670–5674.

12. Gray, C. M., Engel, A. K., König, P., and Singer, W. (1992). Synchronization of oscillatory neuronal responses in cat striate cortex: temporal properties. *Visual Neurosc* 8:337–347

13. Wise, S. P., and Desimone, R. (1988). Behavioral neurophysiology: insights into seeing and grasping. *Science* 242:736–741.

14. Biederman, I. (1987). Recognition-by-components: A theory of human image understanding. *Psychol Rev* 94:115–147.

15. Poggio, T. (1990). A theory of how the brain might work. *Cold Spring Harbor Symposia on Quantitative Biology, The Brain* 55:899–910.

16. Penfield, W. (1975). *The Mystery of the Mind.* Princeton, NJ: Princeton Univ Press.

17. Mumford, D. (1991). On the computational architecture of the neocortex: I. The role of the thalamo-cortical loop. *Biol Cybern* 65:135–145.

18. Baars, B. J., and Newman, J. (In press). A neurobiological interpretation of the Global Workspace theory of consciousness. In: A. Revonsuo and M. Kamppinen (eds.), *Consciousness in Philosophy and Cognitive Neuroscience* (In press). Hilldale, NJ: Erlbaum.

19. Crick, F. H. C. (1984). The function of the thalamic reticular complex: the searchlight hypothesis. *Proc Natl Acad Sci USA* 81:4586–4590.

20. Sherk, H. (1986). The claustrum and the cerebral cortex (Chapter 13). In: E. G. Jones and A. Peters (eds.), *Cerebral Cortex: Sensory-motor Areas and Aspects of Cortical Connectivity* 5 (pp. 467–499). New York: Plenum Press.

A Postscript on Free Will
Pages 265–268

1. Damasio, A. R., and Van Hoesen, G. W. (1983). Emotional disturbances associated with focal lesions of the limbic frontal lobe. In: K. M. Heilman and P. Satz (eds.), *Neuropsychology of Human Emotion.* New York: Guilford Press.

2. Goldberg, G., and Bloom, K. K. (1990). The alien hand sign: localization, laterilization and recovery. *Am J Phys Med Rehabil* 69:228–38.

3. Eccles, J. C. (1989). *Evolution of the Brain: Creation of the Self.* New York: Routledge, Chapman & Hall.

Acknowledgments

Many people have helped me with the writing of this book but a few have made crucial contributions. My colleague Christof Koch, to whom the book is dedicated, has not only collaborated with me in the development of these ideas but has made many detailed comments on the manuscript at several stages. The manuscript has been immensely improved by the trenchant comments of my editor, Barbara Grossman of Scribners. Much superfluous material has been firmly eliminated and what remains has been forcefully edited to improve clarity and ease of reading. That some parts of the book are still not easy to read is my fault, not hers. My personal assistant of sixteen years' standing, Maria Lang, has not only had to decipher my handwriting to produce version after version of chapter after chapter but has also labored over the tedious business of getting the illustrations into the right form and obtaining the permissions for their use, as well as performing all the usual office chores. My special thanks to all three of them.

My thanks also to those who have commented on earlier versions of the manuscript. These include Tom Albright, Patricia Churchland, Paul Churchland, Odile Crick, Antonio Damasio, Peter Dayan, Ray Jackendoff, Graeme Mitchison, Read Montague, Leslie Orgel, Piergiorgio Odifreddi, V. S. Ramachandran (Rama), Paul Rhodes, Terry Sejnowski, and Dan Voll. Their remarks have improved the manuscript and eliminated many errors—they should not be held responsible for those that remain.

I am also grateful to Jamie Simon, who has redrawn many of the illustrations and produced several new ones, often on rather short notice.

Finally, I could not have written the book at all without the loving support and understanding of my wife, Odile, who has put up with my continual preoccupation with these difficult problems over many, many months.

Illustration Credits

Fig. 1. Adapted from Frisby, J., *Seeing.* Oxford: Oxford University Press, 1980.

Fig. 2. Adapted from Kanizsa, G., *Organization in Vision.* New York: Praeger, 1979.

Fig. 3. Adapted from Rock, I., *The Logic of Perception.* Cambridge, MA: MIT Press, 1983.

Fig. 5. Photo courtesy of Becky Cohen, Leucadia, California.

Fig. 8. Adapted from Kanizsa, G., *Organization in Vision.* New York: Praeger, 1979.

Fig. 9. Drawing by Ron James.

Fig. 10. Adapted from a photo by Kaiser Porcelain Ltd. for the Silver Jubilee of Queen Elizabeth II.

Fig. 11. Drawn by Odile Crick.

Fig. 12. Leon D. Harmon, by permission of the Estate of Leon D. Harmon. Photo furnished by E. T. Manning.

Fig. 13. Courtesy of V. S. Ramachandran.

Fig. 14. From Sekuler, R., *Perception.* New York: McGraw-Hill, Inc., 1990. Material is reproduced with permission of McGraw-Hill.

Fig. 17. Copyright © 1988 by the American Psychological Association. Reprinted by permission from Warren, W. H., Jr., Morris, M. W., and Kalish, N. L., *J Exper Psychol: Human Perception and Perform* 14:646–660, 1988.

Fig. 19. Courtesy of V. S. Ramachandran.

Fig. 20. From Julesz, B. J., and Bergen, J. R., "Textons, the fundamental elements in preattentive vision and perception of textures," *The Bell System Technical Journal,* 62(6):1619–1645. Copyright © 1983 AT & T. All rights reserved. Reprinted with permission.

Fig. 21. Adapted from Anne Treisman and Stephen Gormican in *Psychol Rev* 95(1):15–48, 1988.

Fig. 22. Adapted from R. I. Reynolds in *Perception* 10:107–115, 1981. London: Pion.

Fig. 23. Adapted from J. E. Dowling in *Neurons and Networks.* Cambridge, MA: Harvard University Press, 1992.

Fig. 24. From F. Crick and C. Asanuma in *Parallel Distributing Processing: Explorations in the Microstructure of Cognition,* vol 2. D. Rumelhart and J. L. McClelland (eds.), 333–371. Cambridge, MA: MIT Press, 1986.

Fig. 25. Adapted from J. E. Dowling in *Neurons and Networks.* Cambridge, MA: Harvard University Press, 1992.

Fig. 26. From Krech, D., and Crutchfield, R., *Elements of Psychology.* Copyright © 1958 by David Krech and Richard S. Crutchfield. Copyright © 1969, 1974 by Alfred A. Knopf, Inc. Reprinted by permission of the publisher.

Fig. 27. Adapted from J. E. Dowling in *Neurons and Networks*. Cambridge, MA: Harvard University Press, 1992.

Fig. 28. From F. Crick and C. Asanuma in *Parallel Distributing Processing: Explorations in the Microstructure of Cognition*, vol. 2. D. Rumelhart and J. L. McClelland (eds.), 333–371. Cambridge, MA: MIT Press, 1986.

Fig. 30. Drawn by Ramon y Cajal.

Fig. 31. Courtesy of Charles D. Gilbert.

Fig. 32. Courtesy of Charles Stevens from *Sc Am* 241(3):54–65, 1979. Copyright © 1979 by Scientific American, Inc. All rights reserved.

Fig. 33. From F. Crick and C. Asanuma in *Parallel Distributing Processing: Explorations in the Microstructure of Cognition*, Vol. 2. D. Rumelhart and J. L. McClelland (eds.), 333–371. Cambridge, MA: MIT Press, 1986.

Fig. 34. From D. M. D. Landis in *J. Comp. Neurol.* 260:513–525, 1987. Permission granted by the editor-in-chief of *The Journal of Comparative Neurology*.

Fig. 35. Courtesy of Steve Hillyard in *Machinery of the Mind: Data, Theory and Speculations about Higher Brain Function*, E. Roy John (ed), 186–205. Boston: Birkhauser, Inc., 1990.

Fig. 36. Courtesy of Hanna Damasio.

Fig. 37. Courtesy of Hanna Damasio.

Fig. 38. Modified from Eric Mose, "Eye and camera," by George Wald. Copyright © August 1950 by Scientific American, Inc. All rights reserved.

Fig. 39. From *Eye, Brain and Vision* by David H. Hubel. Copyright © 1988 by Scientific American Library. Reprinted with permission of W. H. Freeman and Company.

Fig. 40. Modified from David Hubel and Torsten Wiesel, "Brain mechanisms of vision." Copyright © September 1979 by Scientific American, Inc. All rights reserved.

Fig. 41. From *Eye, Brain and Vision* by David H. Hubel. Copyright © 1988 by Scientific American Library. Reprinted with permission of W. H. Freeman and Company.

Fig. 42. Courtesy of David Hubel and Torsten Weisel.

Fig. 44. Courtesy of David Hubel and Torsten Weisel.

Fig. 45. Adapted from Allman, J., "Evolution of the visual system," in *Progress in Psychobiology and Physiological Psychology*. New York: Academic Press, 1977.

Fig. 46. From LeVay, S., Hubel, D. H., and Wiesel, T. N., "The pattern of ocular dominance columns in macaque visual cortex revealed by a reduced silver stain." *J. Comp Neurol* 159:559, 1975.

Fig. 47. Adapted from Felleman, D. J., and Van Essen D. C., *Cerebral Cortex* 1(1):1–47, 1991.

Fig. 48. Adapted from Felleman, D. J., and Van Essen D. C., *Cerebral Cortex* 1(1):1–47,

Fig. 52. Adapted by David Amaral and Wendy Suzuki from Felleman, D. J., and Van Essen, D.C., *Cerebral Cortex* 1(1):1–47, 1991.

Fig. 56. Adapted from Sejnowski, T., and Rosenberg, C., "Parallel networks that learned pronounced English text," *Complex Systems*: 145–168, 1987.

Fig. 58. Adapted from Duensing, S., and Miller, B., in *Perception* 6:611–613, 1977.

Fig. 59. Drawn by Jamie Simon, The Salk Institute, La Jolla, CA.

Index

About the Author

Francis Crick is the British physicist and biochemist who collaborated with James D. Watson in the discovery of the molecular structure of DNA, for which they received the Nobel Prize in 1962. He is the author of *What Mad Pursuit, Life Itself,* and *Of Molecules and Men.* Dr. Crick lectures widely all over the world to both professional and lay audiences, and is a Distinguished Research Professor at The Salk Institute in La Jolla, California.